GREENING SOCIETY
The Paradigm Shift in Dutch Environmental Politics

Greening Society

The Paradigm Shift in Dutch Environmental Politics

Edited by

Peter P.J. Driessen

Pieter Glasbergen

Professor of Environmental Studies,
Department of Environmental Studies and Policy,
Utrecht University, The Netherlands

SPRINGER-SCIENCE+BUSINESS MEDIA, B.V.

A C.I.P. Catalogue record for this book is available from the Library of Congress.

ISBN 978-90-481-6031-0 ISBN 978-94-015-9958-0 (eBook)
DOI 10.1007/978-94-015-9958-0

Printed on acid-free paper

CONTENTS

Part 4 – Political Institutional Aspects

Part 5 – Conclusions

PREFACE

This book can be regarded as a monograph on the debates and developments in Dutch environmental policy. It has been written with a specific perspective in mind. First and foremost, the line of approach we have taken was from a multidisciplinary social science point of view. The trend in environmental policy is looked at from the angle of sociology, policy studies and political science. Secondly, all analyses depart from the paradigm shift concept. This particular paradigm shift is based on the fact that a radical change has taken shape over the years in the way environmental issues are handled. Previously, environmental policy had always been characterised by is top-down approach in which government determined the actual objectives of policy and assumed that it could win over business, non governmental organisations and citizens to act in line with those objectives. There was also a great deal of faith in the technical solutions to environmental issues. Today's environmental policy is based on a totally different philosophy. Environmental objectives are now reached in association with business, non-governmental organisations and citizens. These actors are also involved in bringing environmental policy into practice. In other words, the implementation of policy has a more interactive nature. New relationships emerge between government, the market and civil society, and policy discourses also become integrated. The environmental interest is more often weighed against the economic interests, the spatial development and against social justice.

The object of this book is both to give an analysis of this paradigm shift and to explore the opportunities and limitations of the new method of governance. The development of the new paradigm is set in the context of further, general social and institutional change. While the analyses relate specifically to Dutch environmental policy, the lessons they learn can also be of significance for the environmental policy pursued in other liberal democratic nations.

This book is the product of intensive teamwork between scientists of Utrecht University (Faculty of Geographical Sciences, Copernicus Institute) and the University of Nijmegen (Nijmegen School of Management). We wish to express our thanks to the authors for the work they have put into this publication and for the inspiring and constructive exchange of ideas that emerged during the process.

Finally, we must also express our gratitude to Rien Rabbers of the KartLab of the Faculty of Geographical Sciences, Utrecht University, for his work on lay out and design.

Peter P.J. Driessen
Pieter Glasbergen

Utrecht University

ABOUT THE AUTHORS

Peter P.J. Driessen is Associate Professor of Environmental Politics and Management at Utrecht University, Department of Environmental Studies and Policy. Most of his research is related to planning and management issues on the national and regional level. He is in particular concerned with the relationships between environmental planning, spatial planning and water management. Furthermore he is engaged in research on opportunities in and the risks of network management.

Pieter Glasbergen is Professor of Environmental Studies holding a chair in Environmental Policy at Utrecht University (Department of Environmental Studies and Policy) and the Dutch Open University. His main interest is in planning and policy issues, particularly with reference to environmental policy, spatial planning, water management, and policy on landscape and nature conservation.

Bas Arts is lecturer in International Environmental Politics at Nijmegen University, Department of Environmental Policy Sciences. His current professional focus is on the role and power of non-state actors in international environmental politics (particularly NGOs and corporations), and the innovation of environmental policy arrangements at the international level.

Carel Dieperink is lecturer in International Environmental Politics at Utrecht University, Department of Environmental Studies and Policy. The main topics of his research activities are environmental policy, international environmental affairs, water management, climate change, policy analysis and evaluation, and environmental innovations.

Henri J.M. Goverde is Associate Professor Public Administration at the Nijmegen School of Management, Nijmegen University, and Professor of Political Science, Wageningen University and Research Center. He chairs the Research Committee 'Political Power' of the International Political Science Association. His research focuses on power, policy networks and multi-level governance. His empirical work concerns policy fields as urban and regional politics, architecture and public space, environment and nature conservation, rural innovation, and water management.

Pieter Leroy is Professor of Political Sciences of the Environment and chairs the Department of Environmental Policy Studies at Nijmegen University. His research and publications focuses on new developments in environmental politics and policies and the emergence of new environmental policy arrangements. He has also a part time professorship in Environmental Policies at Antwerp University, Belgium, where he chairs a research consortium on environmental policies. He chairs the steering committees on Flanders' state of the environment and state of nature reports.

Duncan Liefferink is lecturer in Environmental Politics at Nijmegen University, Department of Environmental Policy Sciences. His main fields of research are European and comparative environmental politics, with a particular interest in the dynamic interrelationship between national and EU environmental policy making. He is co-editor of the 'Issues in Environmental Politics' series, published by Manchester University Press.

Nico J.M. Nelissen was a Professor of Environmental Studies at Tilburg University, and still is a Professor of Public Administration at Nijmegen University. He was a member of the National Council for Environmental Problems and he was also the president of the Dutch Committee for Long-Term Environmental Policy. His scientific work focuses on the responsibility of government for the quality of the natural and urban environment. In this context, he has written several books on environmental consciousness, environmentally-sound behavior, environmental movements, environmental policy, urban renewal, monument care, urban planning, architecture, public administration and governmental policy.

Jan P.M. van Tatenhove is Associate Professor of Public Policy at the Department of Political Sciences at the University of Amsterdam. His research focuses on processes of political modernisation and the institutionalisation and development of policy arrangements, new political spaces and multi-level governance in policy domains, such as environmental policy, nature conservation and spatial planning.

Walter V.J. Vermeulen is a lecturer at Utrecht University, Department of Environmental Studies and Policy. His work focuses on design, implementation and effectiveness of new strategies for increasing eco-efficiency of production and consumption and co-production of sustainable development strategies.

Jochem F.M. van der Waals has been employed since 1997 as a researcher at the Department of Environmental Sciences of Utrecht University. He specialised in environmental policies for cities. Since August 2001 he is employed at the Ministry of Housing, Spatial Planning and the Environment, where he works on traffic policies.

PART 1

INTRODUCTION

Chapter 1

THE PARADIGM SHIFT IN ENVIRONMENTAL POLITICS

Towards a new image of the manageable society

Pieter Glasbergen and Peter P.J. Driessen

1.1 Introduction

Today, environmental politics finds itself in a paradigm shift. Our strategic way of thinking in terms of both content and procedures is undergoing change. The former relates to what we should understand by environmental quality, and how to determine what we should regard as valuable and what we should protect. The latter has a bearing on how to promote that which is regarded as valuable, who should be the change agents, and what rules and structures we need in order to effectively influence societal developments.

If we take a look at the traditional interpretation of environmental politics, we see that both aspects are closely linked within a closed system. Quality is regarded as something that can be determined by science. Whereas scientific knowledge of the physical system could be translated into indicators, against which to examine the actual situation, it leads to a government body establishing a standard which is then translated into regulations and subsequently, by means of supervision and enforcement, imposed on society. In this interpretation, the authorities are regarded as the primary change agents. They must also make use of legal and financial instruments, supported by reliable communication, in order to clarify the choices regarded as essential. The social science knowledge required to bring about change thus follows the scientific definition of the problem. This paradigm was effective for a long time and it enabled the developed industrial societies to rid themselves of the most serious health-threatening

P.P.J. Driessen and P. Glasbergen (eds.), Greening Society, 3–25.

environmental impacts. However, in the meantime doubts have arisen with regard to the functionality of this paradigm (Glasbergen, 1998; 1999).

Today's environmental problems are much more complex and surrounded by much more uncertainty than ever before. There is growing awareness that scientific knowledge regarding the environment is based on disputable estimates of risks. Alongside such environmental risks, account should also be taken on economic risks and arguments of social justice. This is expressed in the liberal interpretation of sustainable development in which quality is interpreted as an open-ended normative, political choice.

Equally important though is that these complex problems are to be handled in a society in which the old, top-down way of governing seems to be on the wane. The management of societal change processes is increasingly occurring in complex networks, in which it is impossible to pinpoint an absolute dominant actor, and in which success and failure are strongly associated with the ability of all parties concerned to form wise alliances and, partly thanks to this, to mobilise their creative potential. The private parties themselves must take responsibility for collective matters; must help to think about what in a certain context at a given point in time are genuine problems, and what are the most appropriate ways to tackle them.

Modern environmental politics defines the issue of guided change as being the problem of establishing a course of action in a collaboration of public and private actors. Government authorities should assure themselves of the cooperation of well-organised private parties who have a direct interest in the issues at stake. Governance therefore needs to take the form of organisation of various discursive learning processes (Glasbergen, 2000).

This paradigm shift is particularly perceptible in western, developed, liberal democratic societies which have formed long-term goals within a perspective of sustainable development. Nevertheless, the process of change has by no means been completed as yet; certainly not in relation to the traditional paradigm, which is simultaneously taken as the basis for effecting the change. This means that it is difficult to define the significance of this new paradigm. It is this significance in particular that is the main subject of this book, in which we analyse the debates and developments connected with the renewal of environmental politics. The developments under way in the Netherlands are in this respect taken as exemplary. With regard to conceptual thinking and new strategy development for environmental politics, the Netherlands – to a certain extent – has the leading edge and it can be seen as an experimental site for the developments and debates on the paradigm shift.

In this chapter we use this new paradigmatic way of thinking to reflect on the political administrative order of today's society. Three key domains can be distinguished in this order: the state, the market and civil society. The pivotal proposition is that in the contemporary environmental politics school of thought a search is made to find new, mutual interrelationships between these domains, as well as to find out how they jointly relate to the physical environment. The physical environment is discussed as the fourth relevant domain: the ecological system. The changing interpretation of this system helps to lay a basis for the paradigm shift. The chapter is concluded with several questions which both authors placed first and foremost for this book.

1.2 The state, the market and civil society

Today's society can be defined as a highly industrial and liberal democratic society. The term 'highly industrial' refers to the stage of development. The society in which the direct health-threatening problems of pollution have slowly become manageable; which to a large extent is technologically capable of manipulating the physical environment; and which is characterised by a well-educated assertive population and a strong economy. The term 'liberal democratic' refers to the basis of the actual order of that society; that which makes it an idiosyncratic society. Lawyers also refer to this society as a democratic constitutional state. This order can be understood in the analytical sense by making a distinction between three domains: the state, the market and civil society. Together, these domains can be regarded as the elements that give it its shape. Each of these domains fulfils a specific function. We shall discuss them here as ideal-typical constructs, in other words, according to their essential characteristics, and the relationships that arise therefrom. Of particular relevance here is the question where the driving force of social dynamism can be located.

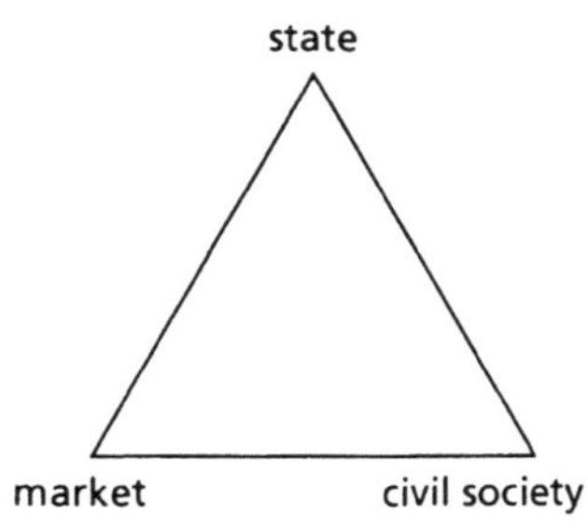

Figure 1.1 Three identifiable domains in society

The state

The state represents the public domain. Its main function is to define, to develop and to implement that which is seen to be in the general interest. Hence the state has a primary responsibility for an orderly society. It creates the right conditions, corrects and it produces whatever is necessary and what cannot be achieved in its absence.

The basis of the state can be defined as political rationality – the organisation of democracy and the political debate, leading to authoritative decision-making processes by representatives appointed for that purpose in society. The *duty* of the state is therefore a task which is mandated by specific rules and procedures. The *authority* of the state is in the latter case always based on power – the ability to achieve certain goals in the general interest by enforcement. Nevertheless, reference made to 'the general interest' as the constituent element for state action indicates that that power may not be used indiscriminately. Both the scope of the function and the implementation are limited.

From a more general point of view, any action taken by the state must first and foremost be functional. Exerting power may never go beyond what society feels is legitimate. Rules must be in line with the sense of justice. This implies a substantial degree of consensus in society. In other words: rules can only be effective if the general feeling is that they reflect the way things ought to be. An increasing number of inflexible rules leads to more efforts being required to enforce them. In a situation of heavy societal opposition, and consequently more intensive control, a regulating government can even endanger its own legitimacy in time.

Secondly, the power of the state must be exercised openly and it must be verifiable, at least in the democratic constitutional state. Moreover, it must be approachable; citizens also have the right to expect the state to account for its actions in exercising its power, and the right to examine such accounts independently. Exercising state power is therefore always slow to transpire and is often reactive by nature. State action is also by definition almost always inflexible to a certain extent. Any changes in these actions will also follow the same slow process. Efficiency is not the predominant criterion when exerting state power. In that respect the process is geared far too much towards administrative care, regulation conformity and legal certainty.

The third limitation on the power of the state is the result of the self-control institutionalised in that power. Checks and balances are built in by the three powers of the state, the division between the legislative power, the executive power and the judiciary power, and also by separate legal responsibilities for national, regional and local administration, as well as territorial and functional-based administrations. State action is always taken within the law and a community's juridical institutions. The function thereof helps to guarantee elementary liberties. The law is thus not only an administrative instrument but it also imposes limits on the administration itself. The law particularly monitors the aspect of one-sidedness in striving to achieve societal goals. In other words, the law generally also limits state action.

Although the fact that the state plays a role in environmental issues is indisputable, the substance and the intensity of that role is set out in greater detail in a process of reasoning. The state bodies that take care of the interests of the environment are also in a weak position in a liberal democratic society. Internally, this is reflected in the many government agencies, all of which often pursue a variety of interests. Those agencies that represent the interests of the environment usually find themselves in a defensive position. If any adverse effects on the environment become visible from other state activities – for instance from economic policy, agricultural policy, policy on traffic – an attempt can be made to take corrective action. Externally, the position is also a weak one in its relationship with society. The important decisions of relevance for the environment – decisions connected with production, consumption and mobility – are taken at numerous locations. There is no specific locus in the decision-making process; it is more often a case of individual decisions being taken on separate aspects of production, technology, etc. Often not even cognisant decisions; everyone is searching for opportunities on highly promising markets. And as long as the relevant procedural rules are followed, everyone is also to a large extent at liberty to make certain choices. Decisions made by entrepreneurs regarding investments are taken mainly in the private domain; the same applies with regard to consumptive choices made by citizens (Glasbergen, 1995, pp. 7-8). The market and civil society therefore each play their own specific role in environmental politics.

The market

The market stands for private interests. We are speaking here about the domain that regulates economic activities and economic production. It is an important motor for prosperity and material progress. The driving force here is economic rationality. The market is competitive, its forces are determined by trading relationships, supply and demand, and it focuses on materialistic objectives.

The essential characteristics of the market make it a perfect or open market. This is a market that meets several criteria (Dubbink, 1999, pp. 50-56):

- it is a decentralised order; there is no authority that coordinates the economic activities;
- coordination occurs spontaneously on the basis of the money-based price mechanism;

- the relationship between the parties on the market is characterised by rivalry and competition;
- the market parties perform a specific role; they seek their own profit;
- private ownership is a constituent of trading on the market; hence there is the opportunity to exercise one's rights on an object;
- the parties are free in their actions; they themselves determine what they produce.

A market with these characteristics knows no other morality than that which is focused on preserving the market itself. This is essential for the function they fulfil, in other words: without these characteristics the function can no longer be fulfilled. These characteristics also determine the link with environmental issues.

The market's first and foremost interest is not to solve environmental issues, although several market parties obviously derive their existence from it. The main aspect is to improve the position in competing for the production of goods and services. While in this respect a long-term strategy can be used, the main concern is still chiefly one of survival in the short term. Environmental degradation, which over the years can have an adverse affect on the market – polluted soil is not very beneficial to farmers – is an effect which under the discipline of the market can be of no other significance than marginal.

The market in today's society is to a large extent dependent on the state. The distinction Van der Straaten makes between the market and market forces is important in this respect. The market is a micro-economic mechanism which is determined by the presence of an adequate number of sellers and buyers. Market forces are macro-economic phenomena which are manifest on the level of the economy as a whole (Van der Straaten, 1995, p. 134). Markets can be, and indeed are state regulated. However, the market forces in liberal democratic societies are forces that must be maintained. These forces are the essence of the domain. In fact, one of the main tasks of the state in this sort of society is to establish and monitor rules that sustain market forces. The state also has a function to perform in those domains where the market comes up against imperfections. This can relate to the deterioration of individual integrity (e.g. the exploitation of children); it can also relate to goods which, while being essential for a society, the market itself is unable to produce. Many forms of environmental quality come under this heading. Nevertheless, because market forces are kept in existence, the overall result of individual decisions on markets are always to a large extent unplanned (Van der Straaten, 1995, p. 134). Accordingly, this also applies to those potential environmental effects which can always emerge from those decisions at a later date; in which case the state can again take corrective action.

For a few, mainly Marxist authors, market forces are reason to assume that – where environmental issues are concerned – the state is a prisoner of those forces. Such authors see the eradication of the logic of capitalism as the only solution. In doing so they deny in fact the independent function of the state and the value of an order which up to now has been the only one to prove its value as the motor for progressive social change. And yet these authors do bring up an important point. In addition to the market's dependence on the state there is a dependency the other way around too. A state which affects the conditions that are essential for a market to function properly must accept a stagnating level of prosperity as the consequence, or in other words: large-scale unemployment. This, in a liberal democratic society, is seen as a failing government policy.

Civil society

Civil society is the domain which is the most difficult to typify. Many people choose to see it simply as a residual category: something which is neither 'state' nor 'market'. However, while it is not primarily occupied with economic rationality, this too is an autonomous private domain. Van Gunsteren defines it as the field of 'the activities and organisations of people with a free choice, free from state coercion and economic gain, and the exclusion of violence in the mutual relationships' (Van Gunsteren, 1999, p. 140). We are talking here about the cultural climate. The actions of the citizen, his standards and his values, his interests, as also the organisations, both political and non-political, that represent those interests without an explicit profit motive.

The function of civil society is two-fold. In the first place, this domain offers personal privacy without external pressure. Within that element of privacy the citizen is only answerable to himself. Secondly, this is the domain in which public opinion is formed. This means that it could also be referred to as the domain of communicative rationality. Civil society is of particular importance for the environment in the second function.

As consumers, citizens are major boosters of environmental problems. Of note in this respect is that whereas the significance of consumer behaviour for environmental quality is often recognised, it is not actually brought into the debate. Influence in the western liberal market economies reaches no further than a potential attempt to make people aware of the adverse environmental effects of consumer actions. It is then a matter of free choice. However, citizens can certainly be influential through the choices they make as consumers. This is not so much an individual capability, but through their joint preferences, organised as an opposing force, they can certainly be an important party in the market. Certainly in relatively prosperous nations, where the market processes are geared more to the demand than to the supply, can they bring about change. One example in this respect were the actions undertaken by Greenpeace which led to Shell being boycotted when it planned to sink the Brent Spar in the ocean after its many years of faithful service. This was reason for Shell to place the aspect of corporate responsibility high on its agenda. A Social Responsible Committee was installed to review the policies and the conduct of Shell businesses. And now, like many other large companies, Shell is specifically seeking dialogue with environmental and consumer organisations. In their joint political preferences, particularly the relative significance they place on protecting the environment, citizens also play a major role as the instigators of change in the actions of the state. The most obvious examples in this respect are the demonstrations against, and the wide public debates on nuclear energy which, in countries such as Sweden and the Netherlands, led to the closing down and later on to the dismantling of nuclear plants.

In other words, civil society's significance for environmental quality is to be seen chiefly through the organisational links of which the citizens are a part. However, they tend to be inconsistent as regards their attitude towards the actual quality of the environment. Some of these organisations foster environmental values (like the environmentalist movement), others express mixed feelings (trade unions, for instance), and yet others put up opposition against meaningful environmental policy, often on aspects which they see as relevant (e.g. automobile associations).

Civil society's relationship with the state is reciprocal. Only the state is able to guarantee that civil society can function freely. This guarantee makes it possible for civil society to serve as the source for determining the nature and intensity of the deployment of state power regarding environmental issues. Recognition of the explicit function of civil society is, for instance, expressed in government funding of the environmentalist movement – as is customary in the Netherlands – under the awareness that this also empowers a major opponent.

1.3 The human domain from a historical perspective

How these domains are positioned is the result of a process of development. They can be traced back to the relationship between the prince, the merchant and the citizen, and then developed further. In certain times, the priest would have also been allocated a crucial role. The church as domain would have been given a prominent place. This domain would have been placed near to the state and civil society would not, or only to a lesser extent, have been recognised as an independent domain. The wider the gap in the relationship between the church and the state, the more likely were religious and spiritual movements to be placed in the emerging civil society. Moreover, they started to play a more marginal role as the energising force in that society.

Another example of change is the recent spectacular shift that took place in the former communist countries in Central and Eastern Europe where we saw the rise and fall of several domains within a short space of time. The unity that had been built up between the state and the market over only a few decades was rigorously violated at the end of the eighties and opportunities have now arisen for the development of what until then had been a repressed civil society. It is quite remarkable to see that in this particular change, civil society played a prominent role in organising the opposition against the regime. A new order was built up out of the remains of civil society – the church being used as a base in Poland. A virtually bankrupt market and a corrupt state legitimated this.

The process of change has not yet been completed, and probably never will be. And so the significance of the state in community life has continued to increase, especially in the 20th Century. Not in the last place because of the government's function in coordinating and facilitating market forces. Moreover, the market has also expanded in terms of scale and is expanding even further to become a global economy. International administrative bodies are also developing which at least supplement the nation state and sometimes replace it. Political parties have lost civil society's interest, making room for new social movements, single-issue groups and informal associations. Environmental organisations have recently formed new international alliances (Arts, 1998).

The roots of social dynamism have also shifted in this reshuffling process. Although each of these domains can function as the energising factor, it would seem that the state is less eligible in this respect. The state prefers to regulate, and is thus able to slow down the dynamism or, conversely, to boost it, but is seldom the initiator. The dynamism comes mainly from the market and civil society, supported by the development of knowledge and technology. The market is an energising force because it is precisely in that domain that the search for new opportunities forms a permanent driving force. Scientific and technological developments have been picked up here and translated into a process that leads from a labour and material-intensive economy to a knowledge-intensive economy. Here, the market itself creates as it were an important precondition for sustainable development: the dematerialised economy. Civil society is an energising force given that a permanent incentive to change is inherent in the actual forming of public opinion, the reflection on and the revaluation of societal developments. The rise of the environmentalist movement, originally in opposition against what were chiefly local developments which were regarded as detrimental to the aspect of liveability, could be reinforced thanks to the fresh knowledge that became available on the background, the magnitude and the impact of environmental deterioration. From the perspective of sustainability, a new energising force developed that placed this knowledge in a normative framework and translated it into change-oriented societal action.

1.4 The vitality of a multiform order

The type of societal order discussed in the foregoing – consisting of three domains – is more or less permanent in all these changes. It has been incessantly tested and in fact its bases have constantly come into view more pronounced. The point emphasised here is the vitality of the multiform order that liberal democratic societies have created. It is this order that should be the point of departure, both in the normative and the empirical sense, in the way we think about environmental politics.

The domains are autonomous in the sense that each independently fulfils an essential function in a democratic, constitutional state. This function can only be fulfilled correctly if independence is guaranteed. It is not by chance that the state is at the top of our triangle. This results from its specific responsibility for the public interest. However, a state which becomes deeply involved in the field of the market creates an inefficient economy. A state which is deeply involved with civil society introduces a totalitarian regime. Conversely, a society that builds up its order on a market basis creates a 'hand for a tooth' society in which the law of the jungle is the only decisive factor. Both market and state can form a threat to democracy. The market because its inherent materialistic orientation tends to reduce all human relationships to bleak relationships of trade. The state because the primacy of violence can lead to a corrupt, omnipotent, internal policy. Each of these destructive forces is unable to keep the other in check on its own. To do this a third domain is required as a counterbalance; the domain of the citizen with the freedom of choice and the freedom of organisation, the domain of the independent debate. Civil society serves to protect the community from the economic power of the market and the political power of the state. Nevertheless, civil society also has its own corrupting forces: the right-wing and left-wing authoritarian movements. In this regard the state forms a counterbalance as the guardian of basic liberties, but the market also acts as a counterbalance which, by stressing such values as private ownership and self-regulation, also gives its implicit support to the autonomy of civil society (Fine & Rai, 1997).

It can be deduced from this normative theoretical underpinning of the triangle that we would like to stress the relative autonomy of the three domains in particular. Only by guaranteeing autonomy are the domains able to fulfil a critical function with regard to one another. This autonomy is relative in the sense that all the domains are simultaneously dependent on one another. The one domain cannot exist without its right to existence being guaranteed by the others. In turn, their functions are also supplementary. Without turning to each other for their supplementary actions, they are all incapable of fulfilling their own function. It is on the basis of these considerations that we do not place modern governance in only one domain, although the state would be the most appropriate one for this purpose, but precisely at the place where the state, the market and civil society, each with its own responsibilities, converge.

1.5 The ecological domain

Sustainable development is concerned not only with human domains but also with qualities of the physical reality. Identified as the fourth domain is the ecological system; this has its own function, a function which has its origins outside the three human domains. The debate on managing sustainable development is mainly concerned with the question of the relationship between these domains. On the one hand, this is a question of how the three human domains relate to one another, and on the other, how they relate jointly to the ecological domain.

The introduction of the ecological system – also referred to as the physical environment or 'nature' – makes us aware of another part of reality. This 'comprehensive' reality exists outside of the three human domains. Indeed, it is the domain from which the other domains derive their very existence. This domain supplies the building blocks for human life, which is often expressed in the two functions that are fulfilled: the source and sink function. The source function refers to the resources the system provides: food, water, air and energy. The sink function relates to the ability to assimilate pollution and waste. Nevertheless, these functions cannot be fulfilled ad libitum. Human influences can endanger them. The question is what significance must be attributed to knowledge of the ecological system and how can we handle that knowledge sensibly. Different viewpoints are involved in the search for an answer, viewpoints which are linked to the recognition of scientific knowledge.

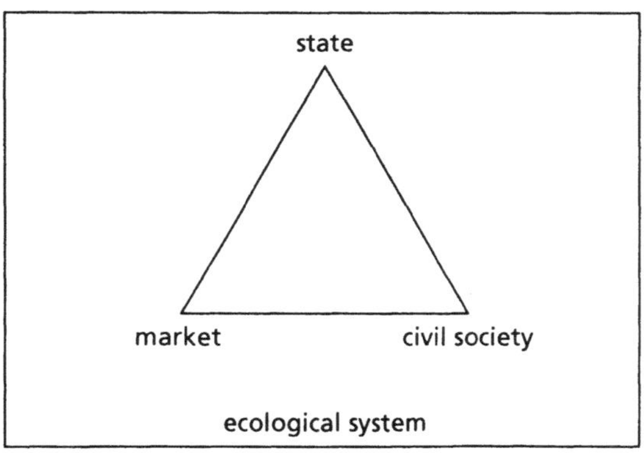

Figure 1.2 The three social domains in relation to the ecological domain

The functioning of the ecological domain is determined by the laws of nature. The characteristics used to give meaning to the domain which are most in line with the inherent characteristics of this system are therefore scientific by nature. The domain is therefore approached as an objective reality that can be identified. Scientific methods can be used in an attempt to bring these laws to light. According to some scientists, this knowledge can be used to determine the pre-conditions within which human activities must remain if the ecological system is to be preserved undisturbed. These limits are expressed in terminology such as 'critical load', 'carrying capacity', or 'ecological utilisation space'. Each of these terms are indicative of a different sort of limit: critical load relates to the limits beyond which pollution leads to a disruption of the ecosystems; carrying capacity relates to the human influences that nature is able to endure; the ecological utilisation space is the widest term – the exhaustion of fossil fuels also attracts attention in this space. The figure below shows the type of thinking behind these terms.

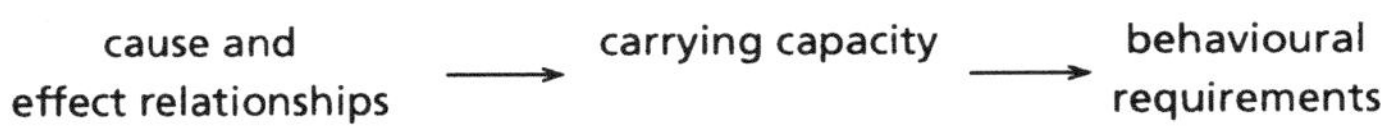

Figure 1.3 Linear thinking

For a few scientists the term 'sustainable development' relates primarily to the ecological system. Instead of evolution they refer to a situation. Sustainability then becomes a situation in which the stock of natural capital is manifest. This is the situation in which the natural reproduction, the regenerative systems, the absorption capacity and other buffers are maintained. In the fundamentalist interpretation, as can be found for example in the interpretations of Hueting & Reijnders, the environmental load the ecological system is no longer able to bear can often be established scientifically with great accuracy. This knowledge calls for political choices to be made, and this, they say, is no easy task because it is a painful process. According to these authors, part of today's consumption will need to be relinquished for example, and the population will need to be reduced. "When making this choice – the choice of an appropriate level of production and consumption – it is not a question of ignoring information on the conditions under which our generation justifies itself in comparison with future generations" (Hueting & Reijnders, 1996, p. 427). In this view, the limits have already been exceeded: in other words the current direction of development is therefore non-sustainable.

Such a point of view is criticised by many. These critics then point out the dynamic character of scientific knowledge – what is true today need not be true tomorrow – and, even more important, the information that is lacking – some of the knowledge required to establish limits is missing and some is uncertain. Knowledge regarding resilience, robustness and persistence is limited, especially for the more complex natural systems. A great deal may be known about parts of these systems and consequently guesses can be made as to the consequences of malfunction. Nevertheless, the consequences for the entire system are almost always the object of speculation (WRR, 1994). The more complex scientific models of ecosystems are, the more unreliable they become. The number of error sources increases too. A good example of this is to be seen in climate models. More research has failed to reduce the uncertainty of the conclusions, but has made them more complex. "A complete ecosystem will not allow itself to be reconstructed in the laboratory", says Van Egmond (1996, p. 14). And yet our inadequate knowledge, which will always be the case, is apparently not the most vulnerable spot among fundamentalists. The main problem lies at the core of this school of thought. We define this problem as the inclusion embedded in the point of view itself. In this context, inclusion means (quite incorrectly) the assumption that the limits that become perceptible within a certain pattern of thought should inevitably also be included in the limits of other patterns of thought.

1.6 Scientific knowledge in the triangle

Here we see the question of the position given to scientific knowledge in the relationship between the state, the market and civil society. If an inclusive position is involved, then that knowledge, particularly scientific knowledge, is the criterion for action in these domains. Inclusivists do have an important point to make in so far as we cannot evade scientific knowledge. It is difficult to find an environmental problem which does not have its roots in one body of scientific research or another. Acid rain, loss of biodiversity, global warming, ozone depletion, desertification and dioxin poisoning are all examples of problems which first began with a set of scientific observations (Hannigan, 1995, p. 75). Exclusion is consequently not an adequate reaction to inclusion. What is under discussion here is how exclusive that knowledge is, and what is the possible function of scientific knowledge in the societal decision-making process. Thinking in terms of inclusion we feel is open to criticism on two points.

First of all the assumption that scientific knowledge is able to function as a unitary whole for policy. Scientific knowledge can only gain significance in a debate in which the value judgements are essential in order to give the facts a context and consequently a societal relevance. To comply with policy relevance, scientific models must simplify reality, reduce the degree of complexity and, occasionally, even ignore certain uncertainties. This is a splendid playing field for other scientists from other schools or with different policy preferences, to bring their claims for knowledge up for discussion in a different context of values. The debate on global warming has for many years been an example of this.

Secondly, the assumption that what in themselves are legitimate claims of scientists, can by definition have one and the same meaning outside their own scientific context. What is important and accurate for the scientific researcher is in the context of policy – even if this claim still stands in peer review – no more than one of the images of the reality that should be included in the considerations. Stakeholders other than scientists can bring up yet other claims for discussion. Policy is by definition developed through dialogue and not something which is deduced from a formal and logical system of thought based on one of the feasible rationalities. Protecting biodiversity is a good example of this. Ethical aspects closely linked to the views on the relationship between man and nature play an important role in this debate. These aspects by definition stay out of reach of scientific positivism.

The function of scientific knowledge stressed here is one of invisible college. This means that science and technology are omnipresent. Scientific knowledge is social, shared by specific groups of actors within the three domains, and occasionally by the domains themselves. In doing so, they not only use that knowledge for their own goals and interests, but they expand it in their day-to-day activities. Knowledge and insight, as it were, being reflected in day-to-day activity: the vehicle of knowledge. This applies with regard to a variety of activities in the market, the state and civil society. This is no reason to trivialise the special expertise of scientists and knowledge of the ecological system. It is reason not to envisage the ecological system as the one and only exclusive basis for determining what a society should regard as sustainable.

1.7 Reflections on the term 'sustainable development'

This point of view can be well discussed by using the model developed by the World Bank's development economists. In addition to the natural capital, which relates to the functions that can be fulfilled by the ecological system, this model also distinguishes three other forms or capital: human, man-made and social capital. This links up three (depending of the authors: four) forms of sustainable development (incl. Daly & Cobb, 1989; Serageldin & Steer, 1994; Goodland, 1995).

The point of departure in this model is the premise that attention should be given to the simultaneous investment in several sorts of capital. Natural capital relates to the stock of environmentally provided assets;

human capital relates to persons in whom investments can be made, for instance in their health, their education, their nutrition; man-made capital is manufactured capital: machinery, buildings, the infrastructure, etc.; social capital refers to the institutional and cultural basis that is necessary to ensure the proper functioning of a society. For the latter sort of capital, legitimacy is an investment option.

This distinction in different sorts of capital is intriguing. A variety of mind games can be played with it. If this is all the human capital there is, the next question is how can we sustain, or increase, the total amount. Another question is whether the different sorts of capital can be substituted for one another, or whether they are complementary. In the former case a reduction in one component, for instance the natural capital, need not necessarily mean a reduction in all the available capital. It is quite possible to become even 'richer'. Playing with the components in this way leads to different forms of sustainability in the model. These are referred to in what is rather suggestive terminology.

'Weak sustainability' is the name given to the form in which the total capital remains intact without the need to be concerned about its composition. In other words, the components can be substituted. This form can lead to the absurd feeling that we might be better off if the natural capital was converted into one or more of the other sorts of capital. 'Sensible sustainability' also maintains the total capital, opens up the possibility of substitution, but stresses the necessity to define critical levels for each component. Here, given that we are usually unaware of where the critical limits are located, it is always a case of being on the alert when substituting the natural capital. 'Strong sustainability' has the intention of maintaining the various components individually. This implies that the loss of natural capital in one area (e.g. a wood that is felled) is substituted by planting the same wooded area at a different location. Oil extraction is allowed, but only if the profits therefrom are invested in sustainable energy. The latter category is referred to as 'absurdly strong sustainability'. The point of departure here is: never deplete anything. Non-renewable resources, such as oil, should therefore be left untouched. Renewable resources only if it concerns the over-mature portion of the stock.

We have somewhat disrespectfully discussed the games that can be played with this model. Although some people have taken the trouble to explain why the one form of sustainability is out of the question, and why the other is feasible, and why they are in favour of the one form and not the other, we do not feel the need to follow suit. This would be inappropriate. As has already been indicated by those working on this model, there is a lack of relevant information. Each sort of capital should be measurable: the production function should be able to be determined in terms of the degree of substitutability and complementarity among the different kinds of capital, dynamically as well; there should be a rate of exchange to allow comparison of the different kinds of capital, and this dynamically as well; and it should also be possible to define sustainability in terms of a context of thresholds within which the more efficient (the highest return) activities could be selected in such a way that individual investments and entire strategies could be meaningfully evaluated (Serageldin & Steer, 1994, p. 31). Also, we are unable to imagine any organisation which has the ability to make these considerations, even if the knowledge is present to do so, and subsequently to take action on grounds of the outcome.

1.8 The merits of the model

For the time being we can only derive more profane conclusions from the model. And yet these conclusions are by no means insignificant. The merits of the model are to be found in the fact that the ecological system is placed in a societal context. We arrive at the following conclusions.

Firstly, that we must constantly make choices within a limited number of investment options. In other words, we are continually engaged in substitution. Not all losses in natural capital is by definition unwise. An example is the extraction of oil, part of the profits from which are subsequently invested in human capital or in exploratory work on other energy resources. Peat was for decades excavated in the Netherlands as an important fuel. Meanwhile there are much better alternatives and peat is no longer of any economic significance. The natural resource of peat has virtually disappeared. While this might be a pity, it fulfilled a very useful function for many years. Knowledge of the scientific limits is therefore in itself less meaningful if we are required to make choices. We must not only look at the negative effects of the destruction of natural capital, but also at the beneficial effects on other concerns of life.

Secondly, that it can sometimes be wise to invest in other kinds of capital, precisely to promote the natural capital at a later stage. Natural capital can also be expanded by way of human actions. In this context Goodland introduced the concept of the 'cultivated natural capital'. We can create woodland and lakes for fish, we can make the land more fertile, improve the quality of food, increase the amount of potable water. In short: "The fact that humanity has the capacity to 'cultivate' natural capital dramatically expands the capacity of natural capital to deliver services" (Goodland, 1995, pp. 14-15). The Netherlands is a perfect example of this. Virtually the entire water economy of the Netherlands is artificial. Thanks to technological developments, converted into new infrastructure, a virtually uninhabitable marshy delta is now a very prosperous part of the world. Obviously a great deal of natural capital has been lost in the process of physical change. Yet there is no doubt that this is progress. We must not only look at the negative aspects of the destruction of natural capital, but also at the positive effects the destruction can have on that capital itself, in other words: the opportunities of transformation inside that capital.

Thirdly, it is important to be aware that in the debate on the significance of the ecological system we do not concern ourselves with a situation in which the paradigm is otherwise quite ambivalent, but with processes. There is a certain dynamism in all the capital components. This is partly determined by society in the sense that social values change. Long & Arnold give the example of appreciation of the marshy region of Florida, the Everglades. "What was once considered progress – for example, damming the Everglades' waterways to reclaim land and promote development – is now considered environmental degradation of a massive order" (Long & Arnold, 1995, p. 21). However, there is also a question of constant movement and a change in the natural system too, just outside the reach of human influence. This dynamism, which is probably more unknown than known, makes it exceptionally difficult to determine what is the wisest thing to do at a given time. While we can take the best possible advantage of this dynamism, it is impossible to give it shape and form on the basis of a concept of a final state which does not exist in terms of natural laws.

This brings us to our final conclusion, and that is that considerations and decision-making processes on sustainable development will always be dependent on the aspects of time and place. A knowledge of the physical system plays a role here, but it can never be decisive on its own. The problem of sustainability is only in abstract a single general problem. The alternatives are different in a situation of poverty than in a situation of enormous prosperity; in a dictatorial regime they differ from those in a democratic regime; in the one geographical system or climate zone they are different than in the other, etc. Again, the Netherlands can serve as an example here too. The entire country has been dug up time and again over the centuries. In terms of biodiversity, this land will not score all that well now. Nature is man-made. This country can be kept an agreeable place to live in by investing more in nature, for instance in recreational space or pleasant, green space. This also occurs for example by allocating farmland as nature reserves, and by realising the essential infrastructure (e.g. locally-controlled flooding of rivers) for developing that land as nature reserves. The preservation of a high level of biodiversity as an international goal must be achieved elsewhere on our planet. There is too little space in the Netherlands to make any real contribution other than behind the scenes. However, considering its economic structure – which is energy intensive – and its wealth, the Netherlands can in principle make a considerable contribution to CO_2 reduction as is focused on in its climate policy.

1.9 Mobilising processes of change

The term 'sustainable development' in this sense steers the discussions on the desired societal development such that it points out the need to confront the different sources of knowledge, and the associated normative views, with one another. In this respect, attention given to the ecological aspects is relatively new. In the sustainability school of thought they are in principle defined as equal to the economic and societal aspects of social development. Where the ecological domain offers choice, the choosing process must be organised – a process which calls for the comparison of considerations concerning the environment, the economy and social justice. This requires a new methodology and the societal organisation of environmental politics.

Instead of *accuracy,* Funtowicz & Ravetz stress *quality* as the criterion for policy relevance. They feel that quality can only be achieved by democratising knowledge; by organising policy-making processes as a dialogue in which a place is set aside for technical and non-professional knowledge, knowledge in the traditional sense and ethical judgements (Funtowicz & Ravetz, 1993, 1994).

This point of view emphasises the necessity for new societal arrangements in which the human domains discussed in the foregoing each fulfil their own role. These arrangements must have specific qualities. It must be possible from each of the human domains to introduce perceptions of risks. Also, the criterion must be imposed on the deliberations that they are organised such that the dominance of certain views is not determined in advance. This implies the need for facilities that give good argumentation a good chance. In this approach to environmental politics we are concerned with the mobilisation of change processes, the utilisation of the potential that is available in society, and the ability to ensure that these processes are permanently kept in motion.

A specific explanation is now given regarding the general problem of governing. Governing in the general sense refers to oriented social change; goal-directed political interventions in society in which the main elements are the wishes or objectives of the intervening public actor(s). Governance refers to a specific process or style of governing in which the point of departure is a plurality of governing actors and a blurring of the lines that separate the public from the private actors.

The governance model is a quite recent administrative model. However, in many modern communities this model is in line with the trend of giving a wide spectrum of societal actors a place in the aim towards achieving sustainable development. We find a powerful argument and a thorough motivation for new arrangements in Lafferty & Meadowcroft (1996) and Meadowcroft (1997). In the first publication they introduce the concept of 'cooperative regimes', indicating that it is quite possible for the representatives of a variety of societal sectors (governments, companies and NGOs) to work together on developing and subsequently implementing innovative solutions to sustainability-related problems. In the second publication, Meadowcroft states: "... it would be possible to establish a vast web of overlapping groups which can explore the practical implications of sustainable development in various dimensions of social life. Some may be based around specific environmental problems or a scarce resource, some may focus on regional re-development, and others may engage with specific industries, products or events. Whatever the point of contact, they will depend on long-term collaborative interaction among representatives of concerned parties" (p. 450). In the more general sense, i.e. not specifically associated with environmental issues, this type of administration was studied and argued by Kooiman (1993) and more recently – but then with a focus on issues of sustainability – by Dubbink (1999).

Giddens' argumentation for a 'third way' is compatible with this school of thought. Nevertheless, his point of view differs in the sense that he focuses on the main ideologies – socialism and neo-liberalism – which he thinks will lose out, leading to the question of which ideological reorientation will need to occur in order to be able to tackle today's problems effectively. Although it is not quite clear whether the new ideology is intended as a reorientation of the social-democratic views or must serve to replace both former ideologies, he too has stopped seeking the solution in unilateral orientation on the collective decision-making of the state or the free market forces, but rather in: "... a process of deepening and widening of democracy. Government can act in partnership with agencies in civil society to foster community renewal and development" (Giddens, 1998, p. 69).

These reorientations on the institutions and ideologies of liberal democratic societies can be regarded as a confirmation of the relevance of trends that are already under way. In today's policy, new social arrangements were recently introduced that should be able to transform the tension between obligations and freedom, public and private – which after all will remain so – into fruitful procedures, aiming towards a sustainable development at this juncture in developed, highly industrialised societies. Several of these arrangements are analysed in the following chapters.

1.10 Implications of the foregoing

Gradually it might have become clear that environmental politics that make sense implies strong human domains which are able to interconnect:

- Each domain fulfils a specific function which can only be realised if they are all relatively independent.

- Because of their relative autonomy, the domains function critically towards one another. This is a prerequisite in a democratic society.
- The domains are interdependent too. Without being assured of the other's existence, and without sustaining one another, they are unable to fulfil their own function.

Strict environmental targets based on scientific evidence is risky if we do not take into account the consequences for the market and the values in civil society. In these cooperations, the three human domains must have the opportunity to bring in their perception of risks. Through deliberations, organised discourses, they can define a common ground for action and eventually share their specific problem-solving capacities. That cooperation should be voluntary in nature, should take place through organised communication, and should lead to contractual agreements (Lafferty & Meadowcroft, 1996; Glasbergen, 1998).

Furthermore, this cooperation can hardly be a one-off activity. As the perception of risks changes we see the necessity to organise cooperation as a learning process; a process in which the three human domains are able to improve their interdependent relations and the common relation they have to the physical environment. The learning process means that the way environmental problems are tackled can be constantly adjusted and thus encourage progressive dynamism.

The observations so far have introduced a new image of the manageable society based on two premises:
- Scientific knowledge regarding the environment is based on disputable estimates of risks. In addition to the environmental risks involved, the economic risks and arguments of social justice must also be taken into consideration.
- There is no better means for sustainable development than the organisation of a cooperative learning process among the representatives of the state, the market and civil society.

This image has far-reaching implications for environmental politics. Complexity and uncertainty are no longer seen as obstacles that can only be surmounted by building up a body of scientific knowledge. Rather, complexity and uncertainty are taken as a challenge to organise the necessary trade-off processes in new institutional relations.

The focus of attention in environmental politics shifts from the building of regulatory and economic incentive structures towards the development of new institutional arrangements to foster collective action (Grabosky, 1995). In the new image of the manageable society, government authorities assure themselves of the cooperation of well-organised private parties that have a direct interest in the issues at stake. They also seek to promote the self-regulating capacity of private parties, stimulating them to tackle issues that are considered collective problems. Private parties are to be assigned a definite role in the public arena. On the one hand they are expected to be active participants in public decision-making. On the other hand, they are to be given responsibility for implementing these decisions.

Key innovative institutions are voluntary environmental agreements, systems of co-management, environmental cooperatives, interactive planning systems, participatory approaches, integrated regional planning and collaborations between private industries and environmental organisations.

1.11 The book in a nutshell

Some societies already work with these new forms of environmental policy. The Netherlands was one of the first countries to adopt the new image of the manageable society as the bottom line of its environmental policy. These changes have certainly not come about without conflict. Moderate conflict is still a part of many of the new arrangements between the state, the market and civil society. However, if we abstract from the policy games and divergent interests that manifest themselves in concrete cases, we see a clear development taking place. Environmental politics increasingly take on the form of organising various learning processes in collaborations of stakeholders in environmental issues.

In the following chapters we analyse this paradigm shift and the discourses that are taking place about it. The development of the new paradigm is placed within the context of more general societal and institutional changes. The point of view taken is one of multidisciplinary social science. It includes elements of cultural history and also elements of sociology, policy studies and political science.

This book can therefore be regarded primarily as a broad social science reflection on the dynamism in the relationship between the state, the market and civil society, the focal point being environmental issues. Special attention is devoted to the role of policy studies in the development of this new paradigm.

This point of view has been translated into a number of key questions:

- What are the characteristics of the paradigm shift in Dutch society, and how can that change be explained?
- Who are the conveyors of the change, both in the conceptual sense and in terms of changing how we think about policy strategies and tools?
- What effect has this paradigm shift had on how environmental issues are approached, and what are the possibilities and limitations that become perceptible in this respect?

– What role does scientific research play in shaping the new paradigm?

The next chapter deals with the conversion of the new paradigmatic school of thought into the network approach to environmental issues. We subsequently reflect on the more strategic innovations in Dutch environmental policy. Part 2 deals with the new environmental politics in the economic context. Attention in this respect focuses on producers and consumers. Part 3 places the spatial context central; a division is made into rural and urban areas. Part 4 deals with several politico-institutional aspects of the new paradigm. The attention in this subject is focused on participation, the new institutional criteria for conducting environmental politics, and the position of the Netherlands in innovations in international environmental politics. The final part of the book deals with the role of policy studies research in the paradigm shift and several conclusions are given.

References

Arts, B. (1998) *The Political Influence of Global NGOs. Case Studies on the Climate and Biodiversity Conventions,* International Books, Utrecht.

Daly, H.E. & J. Cobb (1989) *For the Common Good,* Beacon, Boston.

Dubbink, W. (1999) *Duurzaamheid als patstelling. Over de onvriendelijke betrekkingen tussen openbaar bestuur, markt en civil society,* Eburon, Delft.

Fine, R. & S. Rai (eds.) (1997) *Civil Society. Democratic Perspectives,* Frank Cass, London.

Funtowicz, S.O. & J.R. Ravetz (1993) 'Science for the Post-Normal Age', *Futures,* vol. 25, no. 7, pp. 739-755.

Funtowicz, S.O. & J.R. Ravetz (1994) 'The Worth of a Songbird: Ecological Economics as a Post-Normal Science', *Ecological Economics* 10, pp. 197-207.

Giddens A. (1998) *The Third Way. The Renewal of Social Democracy,* Polity Press, Cambridge.

Glasbergen, P. (ed.) (1995) *Managing Environmental Disputes. Network Management as an Alternative,* Kluwer Academic Publishers, Dordrecht.

Glasbergen, P. (ed.) (1998) *Co-operative Environmental Governance: Public-Private Agreements as a Policy Strategy,* Kluwer Academic Publishers, Dordrecht.

Glasbergen, P. (1999) 'Tailor-Made Environmental Governance. On the Relevance of the Covenanting Process', *European Environment. The Journal of European Environmental Policy,* vol. 9, no. 2, pp. 49-58.

Glasbergen, P. (2000) 'Planning Sustainable Development', In: *Proceedings Euroconference Quality of Live-Sustainability-Environmental Changes,* Event 2, Austrian Research Centers, Seibendorf, pp. 105-120.

Goodland, R. (1995) 'The Concept of Environmental Sustainability', *Annu. Rev. Ecol. Syst.*, 26, pp. 1-24.

Grabosky, P.N. (1995) 'Governing at a Distance: Self-Regulating Green Markets', in: R. Eckersley (ed.) *Markets, the State and the Environment,* Macmillan, pp. 197-228.

Hannigan, J.A. (1995) *Environmental Sociology. A Social Constructionist* Perspective, Routledge, London.

Hueting, R. & L. Reijnders (1996) 'Duurzaamheid is een objectief begrip', *ESB,* 8 mei, pp. 425-427.

Kooiman, J. (ed.) (1993) *Modern Governance. New Government-Society Interactions,* Sage Publications, London.

Lafferty, W.M. & J. Meadowcroft (1996) 'Democracy and the Environment: Prospects for Greater Congruence', in: W.M. Lafferty & J. Meadowcroft (eds.), *Democracy and the Environment; Problems and Prospects,* Edward Elgar, Cheltenham, pp. 256-272.

Long, F.J. & M.B. Arnold (1995) *The Power of Environmental Partnerships,* The Dryden Press.

Meadowcroft, J. (1997) 'Planning for Sustainable Development: Insights from the Literatures of Political Science', *European Journal of Political Research,* 31, pp. 427-454.

Serageldin, I. & A. Steer (1994) 'Epilogue: Expanding the Capital Stock', in: I. Serageldin & A. Steer (eds.) *Making Development Sustainable. From Concepts to Action.* Environmentally Sustainable Development Occasional Paper Series, no. 2, The World Bank, pp. 30-32.

Van der Straaten, J. (1995) 'Environmental Problems from an Economic Perspective', in: P. Glasbergen & A. Blowers (eds.) *Environmental Policy in an International Context, Part 1,* Arnold, London, pp. 133-161.

Van Egmond, N.D. (1996) *Goede raad is duurzaam. Milieukunde en de afweging tussen milieu en economie,* Utrecht.

Van Gunsteren, H. (1999) 'Zwartepieten of selectief sturen? Over de logica van politieke verantwoording', *Bestuurskunde,* no. 4, pp. 136-144.

WWR (Netherlands Scientific Council for Government Policy) (1994) *Duurzame risico's; een blijvend gegeven,* Sdu Uitgeverij, Den Haag.

Chapter 2

NETWORKS AS A NEW CONCEPT FOR GOVERNANCE

Henri J.M. Goverde and Nico J.M. Nelissen

2.1 Introduction

In the past, governance was often envisaged as a series of consecutive, constituent processes (detecting the problem, formulating the goal, the decision-making process, policy implementation and subsequently its evaluation). The policy concerned was then pursued by a government which was seen as (and indeed which saw itself as) the main body responsible for managing societal processes. This representation is referred to as the 'classic model of policy management'. This is a model in which policy is seen as a top-down process. In it, societal actors are regarded as 'objects that can be controlled' and not as 'co-subjects'. Conversely, a society in which the government is responsible for managing new developments together with other parties, calls for a bottom-up approach to governance. This approach was recently labeled as 'managing complex networks' (Kickert et al., 1997).

Over the past few decades there has been an increasing need to deal with environmental issues from the latter perspective (Glasbergen, 1989). In the Netherlands this has led to a 'network approach' (Vos et al., 1994) which is in place at both regional and local level. This approach implies that specific attention is devoted to the field of actors involved in a specific policy, and to the way in which that field can be managed. The latter aspect is referred to as 'network management'. The field of actors is made up of different parties, each having their own perceptions and occasionally even contradicting interests and goals. By pursuing a policy of network management an attempt is made to prompt actors into concerted action. The

P.P.J. Driessen and P. Glasbergen (eds.), Greening Society, 27–45.

network approach should be seen as a paradigmatic shift in terms of how environmental problems are tackled. This approach does not, by the way, depart completely from the classic, top-down views and actions. While network management is moulded by way of horizontal relationships where at all possible, the hierarchical relationships remain important. Both formal and informal relations of power between governments and other societal actors cannot simply be eliminated completely.

The following section (2.2) discusses the main concepts of the network approach. In section 2.3 we briefly set out an environmental problem at the local level which we then analyse on the basis of the network approach. In section 2.4 we take a look at the role played by governments in networks before moving on to look at network management as such, and the different types of network management, in section 2.5. The opportunities, the risks and dilemmas of the network approach are subsequently studied in section 2.6 in relation to the representative democracy and the constitutional state. And last of all, in section 2.7 we put forward several comments on the value of the network approach and the conditions that must be in place to guarantee it's success.

2.2 Policy networks

By thinking in terms of policy networks we attempt to find our way around in the increasing complexity of society, as well as in how to achieve coordination between the parties (Benson, 1978; Godfroij, 1981; Marin & Mayntz, 1991; Benz, Scharpf & Zintl, 1992; Rhodes & Marsh, 1992; Castells, 1996; Goverde & Van Tatenhove, 2000). Societal problems usually have a variety of dynamic aspects and several different facets. The interrelationships between the parties concerned with these problems (governments, market parties and citizens) are also subject to constant change. Societal problems, environmental problems for instance, take on a different character and must be tackled by giving specific thought to each of the problems individually. This results in new forms of management and policy arrangements (Godfroij & Nelissen, 1993). The network approach therefore attempts to detect and tackle (environmental) problems in a more contemporary fashion.

According to Klijn (1996; 1997), the roots of the term 'policy network' are to be found in the interorganisational theory (resource-dependency and coordination mechanisms), in political science (policy communities and sub-systems), and in policy studies (from rational actor to process model). Using these bases, Klijn describes policy networks as more-or-less stable patterns of social relationships between interdependent actors formed around policy problems and/or policy programmes (Klijn, 1997, p. 30).

A policy network exists if the following conditions are met:

- *The presence of several actors,* with a variety of goals, views and interests (multiformity). Hundreds, sometimes even thousands of different parties are involved in virtually every environmental problem (if we regard all individual industrial and farming businesses as parties). Often, not all the parties are actively involved but are represented by umbrella associations and trade organisations. All actors have their own perception of a particular problem, their own goals, their own strategies and their own tactics.

- Actors are interdependent; they must depend on one another to achieve their goals *(interdependence).* Due to the fact that the actions taken by the one party have an effect on the options available to the other parties for them to act, this interdependence differs both in size and intensity. When the government starts to prepare a statutory measure: regulations to minimise the use of crop pesticides for instance, the parties concerned are aware that the regulation can have consequences for their business and that the wisest thing for them to do is to act in advance. On the other hand, if the parties fail to observe certain statutory regulations, then the government is compelled to react (to enforce, to tolerate, or to abolish the regulation).

- At the same time, actors have a certain amount of independence in determining their behaviour *(relative autonomy).* They devise their own strategy in the light of the goal they aim to achieve (action logic). Different interests are also often concealed behind the various goals, perceptions and such like. Whether a party chooses for relative autonomy, or for a position which is more allied to that of others, depends on the sources of power available to that party. Some parties derive power from statutory rights, others have large sums of money or knowledge at their disposal, while others have a considerably large, easily mobilised group of followers. The mutual relationships within the policy network are determined to a large extent by the power factor. And yet for policy networks in general, no single party can run the show on his own. Coalitions can arise within a network which are focused, *inter alia,* on the dominant position of a certain actor. Some parties single one another out in order to gain strength and thus achieve their own goals.

- Processes of interaction among the actors are characterised by complex exchange relationships *(interweavement).* A divergence in interests leads to strategic behaviour among the parties: they become alert, are on their guard for one another, try to keep one step ahead, contact one another, provoke a conflict if necessary, avoid risks, or show indifference.

- The relationships and interaction between the actors often display set patterns which over a period of time become more or less permanent *(sustainability)*. There is also a considerable level of *dynamism,* among other things due to processes of actor exclusion and inclusion and the associated cultural changes (discourse changes: changes in the predominant norms, values, rules, symbols and language). How a policy network develops over time is never absolutely clear. There are far too many parties and too many nuances in the mutual relationships to be able to predict precisely how a policy network will act. Not only external circumstances lead to dynamism in policy networks; policy networks also have a dynamism of their own. Minute, internal shifts give rise to processes which in due course can lead to an entirely different pattern of the policy network.

Policy networks are usually formed around specific policy problems (manure-related pollution, pollution from CO_2 emissions, surface water pollution, waste processing). Occasionally, policy networks arise around problems that go beyond the policy community (the environment and the economy, the diminishing biodiversity, climate change). In other words, there are many different types of policy networks. There are as many policy networks as there are environmental problems. It is even conceivable for there to be several policy networks for a single environmental problem.

Policy networks fulfil a variety of different functions for the participating actors. Which functions depends, *inter alia,* on the nature of the interdependence in the network. If it is a *symbiotic interdependence,* i.e. the actors are able to fulfil their goals by exchanging resources, then the actors will generally be driven towards one another. They are then complementary; coordination is possible. Things are more complex in the case of *competitive interdependence* where the actors have a need for the same resources, i.e. they get in each other's way. This has a repellent effect; either that or the one takes over the other in order to rid himself of a competitor.

2.3 'Schuurfeesten': an environmental problem in a complex local context

The network approach calls for a different way of thinking and acting than was customary in the classic management model. The following example will illustrate this.

Case description

A farmer has had a new barn built in which to house his pigs. Given that the government is constantly imposing more statutory environmental requirements which are becoming even more strict (e.g. for the purpose of combating the emission of ammonia), he has stopped using the barn for his pigs but now uses it as 'storage for caravans' (usually owned by city-dwellers) and as a venue for what is a typically Dutch rural phenomenon, 'schuurfeesten', or 'barn parties'. Hundreds of youngsters from several surrounding villages attend these parties on Saturday nights. The villagers (a population of about 10,000) are pleased that the youth has the opportunity to run riot outside the village. However, according to the neighbours, people from the city living at a distance of some 900 metres from the barn and separated by open pastureland only, they cause a great deal of (noise) nuisance. These neighbours took the matter to court a year ago. Since then, they themselves have been intimidated and their property vandalised. Reason enough for them to frequently 'go into hiding', or in other words, they feel forced to retreat to a secret address. The pig farmer's property also borders on an estate which has been owned for some 300 years by persons of national standing (the former mayor of Amsterdam, MPs and the Minister of European Affairs in a former cabinet). While the owners of the estate did register a protest against the license for these 'barn parties' they did not take the farmer to court. The owners' interests are represented by a lesser well-known brother who lives in a village some 60 km away. This brother acts through his lawyer only.

The court has meanwhile ruled that the 'barn parties' are not in conformity with the council's policy on spatial planning. The municipal authorities subsequently decided that only three more of these 'barn parties' may be held this year. The result was a protest on the part of the entrepreneur as well as among the youth. A sort of people's tribunal has even been set up. Unknown persons have used pickaxes to cause havoc on the neighbour's land. Thirty tonnes of demolished property had to be removed. In an interview with the local newspaper, the mayor states that the neighbour is formally in the right. In other words, said the mayor, the neighbour did not act all that cleverly. He would have been much better to have sat down with the farmer at his kitchen table and had a good talk, instead of taking the matter to court or accusing the local council. He said: "There is only a 'minimum of nuisance'. These parties have been tolerated for almost 20 years now. The barn is located at a considerable distance from the neighbours. If an outsider moves into the vicinity and is unable to agree with how things are done here, then he will soon be faced with problems. These acts of vandalism were certainly not carried out by the youngsters of the village in question." According to the lawyer of the owners of the estate, the mayor is giving an incorrect representation of the matter: "It is the municipal authorities that were unwilling to enter into discussion." For that matter: "because of the ties with the district, the family had adopted a reticent attitude". A former mayor of a neighbouring municipality made an informal attempt to mediate in the matter, but failed. According to this 'mediator' the owners of the estate will probably now also take the matter to court.

Yet there is more background information on this story. While it is true that these parties have been held for almost twenty years now, originally they were only held on New Year's Eve. The municipal authorities were happy with this: "You must give the youth the opportunity to go out and party or you get your bus shelters and telephone booths vandalised." These 'barn parties' soon became popular. The Regional Farmers Youth Association gave nine of these parties a year which were attended by some 1,000 youths. Other activities were added to the list as time went by: tractor-pulling competitions and family parties. Especially the rise in the actual number of activities was the reason for the (new) neighbours to go to court. In the early 1990s the municipal authorities informed them in writing that the parties were not in conformity with the planning regulations and that the farmer could incur a fine of € 25,000 per occurrence. Nevertheless, the authorities have never imposed a fine. This was because policy had been changed; a report had shown that there was too little opportunity for nightlife activities for the youth in the municipality.

The farmer relishes this. He sneers about the fines that had been announced, and sees himself as a trend-setter who had helped the local community to channel violence, nuisance and crime among the youngsters in the surrounding villages. He responds laconically to the commotion: the barn had been built for pigs but government legislation had prevented him from using it as such. Quite impertinently he adds: "otherwise I would have had pigs every day, now I only have them once a month".

Case analysis

This case illustrates that there are several 'worlds' involved in any problem of policy. A distinction can be made between the 'established', the world of the local, rural culture (the farmer, the youth and the local authorities) and the 'outsiders', the culture of supra-local regimes (the legal powers, the political elite) and lifestyles (urbanites in the countryside). Alliances are occasionally formed between these separate worlds. For instance: the pig farmer who does not populate his barn with pigs (and therefore does not contribute towards the unwelcome increase in number of livestock) but makes it available for 'barn parties', is an entrepreneur who generates a new economic pillar in the countryside.

The 'barn party' phenomenon is apparently important for the identity of rural youngsters. These events differ enormously in terms of culture from one rural area to another. An anti-urban mentality is consciously cultivated in several rural areas, and in this respect, the enterprising farmer helps maintain the norms and values of the district. Local councils see these parties as a good means of efficiently enforcing public order. Any nuisance is then concentrated, and kept outside the village centre, implying that little public property (street furniture) is vandalised. The negative aspect is that gatherings of this kind can always cause a considerable amount of nuisance (noise, crime, quarrels) to private individuals. Nevertheless, you should not remedy this by laying down rules in planning regulations. Sensible enforcement of public order calls for skills in certain, locally-oriented protocols (the use of codes, rituals and symbols): an informal talk over drinks is more effective than a civil code. The owners of the estate, as 'long-standing outsiders' but 'good neighbours', are much more aware of this than the 'new country folk from the city'. Moreover, they are probably dependant on a good relationship with several members of the local population for the upkeep of the estate.

At first glance, the problem would seem to be one of a clash between a 'clever' farmer who knows how to survive in this rural community, and an assertive 'neighbour' who brings the institutional dimension of his problem into play. The environmental aspect of the problem is apparently limited to the aspect of noise nuisance in a quarrel among neighbours. However, if we take a deeper look we see that the environmental problem is part of a contested space in a rural community. While both neighbours apparently represent different worlds, they are also interdependent. In fact there are very many actors involved and, moreover, all participants are apparently obsessed by their own interests. The archetypal instruments for resolving this conflict are evidently ineffective. Neither juridical (rules, licenses, the courts, toleration) nor communicative (statements published in the press, mediation) instruments seem to work. The imposition of economic instruments (subsidies, levies) was not evident.

Management strategies in networks

What is the next step in this local environmental issue, embedded in social and political relationships and various local rural cultures, as well as in supra-local and local institutions? Network management teaches us that there are two ways of reaching a solution: a) a variation in the number of participants; b) a variation in the interaction among the actors. Both dimensions can focus on either improving the game in the network, or on a total reconstruction of the network as a whole. According to Klijn & Teisman (1977, p. 106) the following management strategies can be distinguished in networks:

	perceptions	actors	institutions
game management to improve game	covenanting: exploring the similarities and differences in actors' perceptions and the opportunities that exist for goal convergence	selective (de)activation: (de)mobilising actors who possess resources (to block a game)	arranging: creating, changing and sustaining *ad hoc* provisions which suit groups of interactions
network constitutional changes to improve the network	reframing: changing actors' perceptions of the network (which games to play, which professional values matter, etc.)	network (de)activation: bringing in new actors, or changing the positions of existing actors	constitutional reform: changing rules and resources in networks or trying to fundamentally change the ecology of games

Table 2.1 Management strategies in networks (Klijn & Teisman, 1997, p. 106)

The mayor, the local government's representative who is usually also responsible for the public order and, as such, holds a co-supervisory position regarding police action, is unable to manifest himself in such conflicts simply as 'one of the many parties' (a horizontal relationship). While a hierarchical leadership is also expected of the mayor, his actions to date have afforded him little opportunity to use means of communication in order to investigate the chances of goal convergence (covenanting), or to deactivate certain actors, especially the 'neighbour'. And yet he can try to establish contact with the youngsters to persuade them not to cause a nuisance at the remaining 'barn parties'. Making convincing threats to take police action is included here (arranging; bringing in new actors), now that the court has delivered judgement (constitutional reform). Via the press he can also attempt to change the way the actors perceive the situation (reframing). The enterprising farmer could do the same thing. The village's municipal council could also change 'the ecology of the game' by announcing a change in the planning regulations, focusing on a more regulated authorisation of the 'barn parties' (causing no noise or social nuisance), emphasising the general interest in the longer term. The assertive 'neighbour' could be given the prospects of damages and/or a subsidised removal.

2.4 The government and policy networks

In the case of environmental issues we frequently see a gap between the ambitions and the actual results achieved. The traditional form of management, in which a problem is regarded as a technical issue that can be steered on the basis of goals drawn up centrally by government, overestimates the options for government control while at the same time underestimating the multiformity of interests. The somewhat ineffectiveness of the environmental policy pursued to date can be reduced to the aforementioned overestimation and underestimation. Network management is for the purpose of starting up a process of interaction and communication in which the actors reach agreement on how to tackle a complex environmental issue. We can deduce from the case in question that basically all the participants act in such a way as to make sure that the network functions as well as possible in accordance with their own action-logic. The government has become very much aware of this over the past few years.

Driessen et al. (1993, p. 146) conclude from an analysis of the environmental policy network in the Gelderse Vallei *(Gelre Valley)* and in Zeeuwsch-Vlaanderen *(Zealand Flanders Canal Region)* that the process of consensus forming is more effective if:

- it can be demonstrated that the continuation of the existing situation is not an attractive proposition for any of the parties involved;
- it is clear that the parties concerned cannot break through the stalemate without help and bring the solution to the environmental problem closer;
- a structured process of interaction and communication can be started up between the most involved public and private parties, and thus bring the environmental goals into a wider perspective of development for the whole region;
- this process leads to a regional view on development, which those concerned see as a total result of the negotiations, and in which everyone has more to gain than lose in due course.

The government is aware of its inability to solve the environmental problems unilaterally. Moreover, it is in line with the new public management approach that the government takes a backseat in favour of the market parties, and that governments should not be too predominantly present in policy networks (Nelissen, 1994). It is therefore typical of network management that the government relinquishes its hegemonic position and acts just as one of the many parties where at all possible. The consequence is that the government is now entering into all sorts of public-private partnerships (co-production) in order to be more successful.

However, if the interweave between public and private organisations becomes so intense, it could be at the expense of government's credibility and legitimacy.

Nevertheless, the government remains one of society's special actors. After all, the government's first and foremost task is to constantly guarantee the constitutional rights of its citizens. The selective activation and exclusion of actors in network management is, for instance, contrary to the constitutional prohibition regarding discrimination. Secondly, governments are called upon by citizens and organisations to actually implement certain tasks which are regulated by law, if necessary against the wishes of the participants in a certain policy network. Thirdly, it is the government's democratic duty to take into account the interests of parties who do not belong to a policy network. And finally, the government has the monopoly on means of defence, i.e. control over the potential deployment of the armed forces, the policy and the judiciary as the instruments over which other parties should not have at their disposal (De Bruijn & Ringeling, 1997, pp. 153-154).

In the main it can be stated that the government fulfils the roles set out below, particularly with regard to the environment:

- it lays down the environmental norms (in this case: the relevant target values);
- it creates structures for consultation and negotiation frameworks;
- it creates juridical, financial and economic and communicative conditions;
- it coordinates and monitors progress.

In terms of policy, it is the government's duty to establish target values for the quality of the environment. Among other things this relates to emission values such as the values for SO_2, NO_x, CO_2 and NH_3. As a rule, these target values are coupled to a certain period, for instance: reduce the emission of CO_2 by 10% in the year 2010 as against the emission figures in 1990. Moreover, the government sometimes promotes the formation of policy networks on environmental problems (network constitution). In this context it is appropriate that the government helps to set up consultation structures and frameworks for negotiation, as well as helping to structure the contacts between the various parties (relational network management). Another role of the government is to create favourable conditions for achieving the collective goals, for instance by facilitating the availability of funding and expertise (research institutes, management consultancies). Also, if the set target values and dates are to be met, the government must often assume responsibility for coordination, among other things by promoting support and taking responsibility for monitoring the progress of a policy process.

2.5 The management of policy networks on environmental problems

Working on the basis of a knowledge of the structure, the workings and the dynamism of policy networks, is it possible to find a pretext for the management thereof? In order to answer this question, this section takes an in-depth look at network management.

What is policy network management?
De Bruijn et al., (1993, p.23) define network management as: "The way in which actors make a deliberate attempt to influence the structure, the workings of and/or the policy outcomes of a policy network". The general view is that network management is not aimed at achieving pre-determined policy outcomes, but at achieving collective solutions to problems and at ensuring good interaction or good relationships among the actors. In their publication on network management, De Bruijn & Ten Heuvelhof (1999, p. 56, 144) state that network management is concerned with developing a redundant (superfluous) relationship pattern (relationship management). Redundancy in relationships implies that information comes in from various directions and that it is less easy to establish a power monopoly. The participants are consequently less obsessed with their own goals and interests and are more willing to place their aims to achieve their goals in a longer time frame; in short, there is a willingness to be more flexible. In this way, the actors have several alternatives for their actions. In other words, they have more room in which to manoeuvre.

Who manages the policy networks?
Network management presumes that governments do not have the monopoly on network management. The role of the 'network manager' can also be fulfilled by other parties. In other words, the government itself will be the 'object of management' for other parties. The 'network manager' is by no means an 'administrative heavyweight' who stands way above the others. In fact all participants in a network are their own chairmen: working on the basis of their own action-logic they attempt to influence the actors and the mutual patterns of interaction to such an extent that the network as a whole functions more in line with their own goals and interests. It goes without saying that not all actors in the network are able to control the progress of the process to the same extent. Depending on the intensity of his interests, the control he has over the means of exercising power, and the extent of his access to (formal) decision-makers, the one actor has greater or lesser management abilities on the network as a whole. With a view to the position held by governments in several environmental policy networks (tending to the general interest, having ample means of exercising power,

and often being the formal decision-maker) it is quite natural that they have a powerful position in the network. At the same time, the mutual dependency among the actors means that none of them are in a position to get their own way to the full extent in the network on their own.

The different kinds of policy network management

Management of policy networks can be structured in different ways. First of all we can distinguish between subject, structure and process management. Termeer (1993) also distinguishes administration, intervention and trajectory management.

Subject management is concerned with promoting a certain solution for a policy problem, for instance a reduction in the number of acid equivalents. Suppose that the government acts as a network manager, then – as one of the many parties – it is able to promote that the interaction between the parties is focused on achieving this goal. It goes without saying that the government's special status demands that it makes an attempt to influence a policy network on grounds of subject by means of making a vertical contribution (e.g. by establishing norms in regulations).

The view generally taken is that only structures and processes are managed in policy networks. *Structure management* focuses on such matters as entrance to (inclusion) or expulsion (exclusion) from actor policy networks, thus increasing or reducing the number of participants accordingly. Also, aided by regulating the mutual relationships in the network, the multiformity or the degree of representation of actors is either increased or lessened. For instance, central government can change the number of members in an environmental advisory council.

The *process management* of policy networks relates to steering the course taken in policy processes. De Bruijn et al., (1998, p. 70) documented experiences in this field and developed design principles. Process management is important for several reasons. A reduction in the level of uncertainty regarding the subject matter is often achieved. As a rule, the actual definitions of and the solutions to the problems are enhanced. Decision-making generally becomes more transparent and a certain amount of depoliticising takes place in the decision-making process. The ultimate outcome is more support, leading to less fear of a defective coordination of policy and laborious implementation.

Three other forms of network management were worked out in the configuration approach; an approach developed by Termeer (1993, pp. 278-299) for the purpose of analysing the inertia when solving the problem of manure in the pig farming sector: administration, intervention and trajectory management. Each form of management has several strategic options that can be used to benefit the managers of policy networks (see also Table 2.1).

The strategy of *administration* focuses on facilitating processes of continued interaction. This strategy relates to both the 'what' (definitions of reality) and the 'who and how' (relationships). It is a strategy that relates *inter alia* to the following activities (Voogd, 1990):

- creating a common language: pig farmers and environmental activists, for example, speak such a fundamentally different language that they are unable to understand each other;
- facilitating the elements of 'play' (the process of establishing and changing the rules of interaction) and 'game' (the process of constructing and reconstructing the definitions of reality);
- guarding the access of 'third parties'; for 'whom', for 'what' or 'how' do actors split off in a policy network?;
- promoting contemplation: preventing fixations and consequently blocking interaction requires that the actors show a willingness to contemplate (i.e. to bring their own definition of reality up for discussion);
- warding off discord in functional conflicts: if the discord between realities eventually leads to converging reality definitions, then the conflict has been functional.

The second strategy is that of *intervention*: breaking through blockades in interactions. This can be achieved by means of 'context variation', a method that can be worked out by following two different techniques, i.e. by introducing 'third parties' (new actors, new rules of interaction or new definitions of reality) or by 'reframing' (changing the organisation by encouraging or enforcing contemplation). The latter technique relates to a radical, suddenly changed situation of perceptions and behaviour, resulting in a change in the significance of a certain policy strategy, as well as the associated consequences, while the facts as such remain unchanged.

The third strategy, *trajectory management,* is intended to influence the substantial outcome in particular. Trajectory management calls for intervention in a dynamic process at a given point in time. This is more feasible in a change trajectory which has not yet been completed, as opposed to when a completely new initiative in terms of content has to be introduced. Moreover, this type of management is easier if tentative statements can be made as to the various development possibilities, thus eliminating any influences that can lead to a move in a direction which is not in line with the trajectory already under way. This does not however mean that it is known which trajectory is the most desirable one – considering the potential developments – and whether the process can be encouraged in that direction. This strategy is more likely to be one of making constant adjustments than one of making plans in advance as to how a required trajectory can be achieved. This is why Termeer does not think the chance of this happening is very high.

2.6 The opportunities and risks of network management in a representative democracy

Several normative aspects are attached to thinking in terms of policy networks and network management. These normative opportunities and risks should not be neglected, certainly not those in the field of environmental policy.

The opportunities involved in network management relate mainly to widening the responsibilities for societal solutions to problems, the opportunity to experiment (policy networks are not subjected to a statutory setting), the problem-oriented character and the strong linkage between the preparation and implementation of policy (Driessen et al., 1995, pp. 84-88). Network management has a major mobilising effect. The parties, including those parties that initially opposed one another or adopted a 'wait and see' attitude, are brought into action. The often existing, inflexible mental boundaries are released and we then see creative solutions emerge.

Risks evolving from a democratic and constitutional state perspective which are associated with network management are also pointed out (Nelissen, 1993). Not all the outcomes generated in policy networks are able to follow in the footsteps of formally established policy. That could affect the legal certainty of 'third parties'. Citizens will sometimes wish to lodge a complaint against decisions taken in policy networks, yet it is not always evident against whom that complaint should be made. It is quite possible that the outcome of a win-win situation can be achieved within policy networks for the parties concerned, but that the products or network management can occasionally have (adverse) effects on groups that are not a part of the network, or are unable to influence the policy product and/or its realisation. The consequence can be one that the products are not, or inadequately, accepted by society. The problem-oriented character (issue networks) of network management increases the chance of more integral solutions. At the same time, the aspect of selectivity regarding what is and what is not taken into consideration cannot be excluded. The linking of preparatory work to implementation implies that the parties will be morally tied to the consensus that has been achieved, and that the individual responsibilities for implementation can be emphasised. On the other hand, this linkage also implies the risk that the policy network will claim too much of the authority that belongs to the representative bodies. If the results of processes cannot be tested by means of network management, then this will be a source of opposition, certainly in combination with the selective recruitment of actors in a policy network. Another major risk is in the representation of the representatives of interest groups in policy networks. Are they able to make decisions themselves or must they always consult their followers? There must always be time incorporated for consultation with the followers.

By setting out the risks of network management we have already indicated that managing (environmental) policy networks calls for constant, critical contemplation. The interweave between governments and societal actors has always been very closely knit in the Netherlands, at least since 1945. This was originally referred to as the 'neo-corporativist decision-making process' (for a focus on agriculture, see Chapter 6). This method of decision-making was based on intertwinement between governments, employers' organisations and the unions in a variety of branches of trade and industry. Influenced by the network approach however, the interweave characteristics have changed. The policy process is difficult to imagine as being divided into separate stages (policy preparation, consultation, the decision-making process, policy implementation, evaluation). The former interweave was found mainly in the stages of consultation and implementation. Conversely, pursuing policy in policy networks is a process which is less easy to identify, in which those involved have become less approachable and less verifiable. Accountability – by whom and to whom? – has also become less transparent. Another major characteristic of policy networks is their so-called *'Eigengesetzlichkeit'*: the further the process progresses, the more they become entities sui generis; they take on their own structure and dynamism. It is for this reason that external control over networks is not easy to achieve. Policy networks can thus adopt the attitude that they more or less represent the legitimate power in a certain area of policy, and that they consequently speak and act on behalf of the persons, parties or social groups concerned. This gives them a 'semi-chosen' arrogance and they subsequently regard control as unauthorised intervention in their own affairs.

Members of parliament often have difficulty in accepting 'tailor-made management' (Vos et al., 1994) through policy networks. Parliament consequently fears that 'the primacy of politics' would be taken too much out of their hands if management were to take place via networks. We also frequently see hesitation on the part of the environmentalist movement regarding the new approach. Nevertheless, the parties involved in complex environmental problems are still dependent on one another, and there is often no other solution possible than to bring a solution closer through network management.

2.7 Conclusions

The perspective of policy networks has become very popular over the past few years. Network management is a major exponent of the change in relationships between governments, NGOs and citizens over the past twenty years. From the perspective of rethinking environmental governance the network approach seems to support some post-modern characteristics in society. "In the place of the static individualism of liberalism and the oppressive collectivism of Marxism, post-modernism stresses fragmentation, relativism and multiple, often contradictory, identities" (Faulks, 1999, p. 166). On the one hand the network approach privileges not a particular identity, one special fragment or a 'specific thruth'. In this way the role of the state has become less dominant in relation to civil society. That is why the institutions of government do not have the administrative capacity to solve environmental problems on their own. They are mutually dependent on other actors in society. In this line of argument, although they cannot be labeled as pure post-modern authors, the proposals made by Giddens (1994; dialogic democracy) and Beck (1997; sub-politics: the politisation of civil society) for the development of social order as a key issue of governance seem to be realistic under the characteristics of a 'network society' (multiformity, interdependency, relative autonomy, interconnectedness, sustainability; section 2.2).

On the other hand, the strength and impact of environmental problems can not be approached successfully from a (post modern) perspective, including an extreme and nihilistic relativism that leads to a pre-modern fatalism or to the permanent triumph of the strong over the weak. Governance based on a radical liberterian view, where all that matters is the freedom to choose and not the nature or the consequences of the choice itself, leads to a dead end as well. Environmental problems need state institutions which are more than a manifestation of power relations that are all around us and cannot be properly located, as well as a civil society that is more than a market place in which we engage in a multitude of life-style experiences, with little apparant obligation to anyone, or anything else (to paraphrase Faulks, 1999, p. 167). In fact, network steering and network management attempt to offer an institutional framework that provides the state with the opportunity to set the main policy-goals in a reference polity, and civil society with the opportunity to arrange a public space that produces consensus about innovative policy-solutions.

And yet if we look at the summary of the risks and at our reflections on the appropriateness of policy networks in the system of parliamentary democracy we still see many dilemmas relating to the network perspective. Hence there are two methodological considerations about network-management that should be emphasized:

- network management is an adequate approach on condition that participation is voluntary; the problems are complex; the interdependency is substantial; there is no situation that poses a catastrophic threat to the public health; and that the status quo must be changed (Driessen et al., 1995, p. 90).
- networks must be regarded less on the basis of 'deciding' and more on the basis of 'interaction and playing'. This implies that particularly the government should regard networks less as a modern instrument (network constitution) they can use to take more effective action then ever before (network management). Explained as a 'game', the network perspective offers a more realistic view of the policy processes (patterns of interaction, balances of power). Moreover, it creates the opportunity to bring about learning processes in such a way that (environmental) problems will be able to be solved in due course.

Finally, it can be concluded that network management functions as a new concept of governance within the changing relations between state and civil society. However, the main characteristics of networks can easily promote a political relativism and a neo-liberal polity oriented on life-styles and fashions only. In such a setting, the network approach will not be able to legitimately tackle environmental problems. That is why network management as a new concept of governance requires state institutions that give guidance and a civil society that arranges environmental innovations within a responsive and accountable polity.

References

Beck, U. (1997) *The Reinvention of Politics,* Polity Press, Cambridge.

Benson, J.K. (1978) 'The Interorganizational Network as a Political Economy', in: L. Karpik (ed.) *Organization and Environment,* Sage Publications, London, pp. 69-102.

Benz, A., F.W. Scharpf & R. Zintl (1992) *Horizontale Politikverflechtung. Zur Theorie von Verhandlungssystemen,* Campus Verlag, Köln.

Castells, M. (1996) *The Rise of the Network Society,* Blackwell Publishers, Malden/Oxford.

De Bruijn, J.A., W.J.M. Kickert & J.F.M. Koppenjan (1993) 'Inleiding: beleidsnetwerken en overheidssturing, in: J.F.M. Koppenjan, J.A. de Bruijn & W.J.M. Kickert (red.), *Netwerkmanagement in het openbaar bestuur. Over de mogelijkheden van overheidssturing in beleidsnetwerken,* VUGA Uitgeverij, Den Haag, pp. 11-30.

De Bruijn, J.A. & A.B Ringeling (1997) 'Normative Notes: Perspectives on Networks', in: W.J.M. Kickert, E-H. Klijn & J.F.M. Koppenjan *Managing Complex Networks. Strategies for the Public Sector,* Sage Publications, London, pp. 152-165.

De Bruijn, J.A., E.F. ten Heuvelhof & R.J. in 't Veld (1998) *Procesmanagement. Over procesontwerp en besluitvorming,* Academic Service, Schoonhoven.

De Bruijn, J.A. & E.F. ten Heuvelhof (1999) *Netwerkmanagement. Strategieën, instrumenten en normen,* Lemma, Utrecht.

Driessen, P.P.J., P. Glasbergen & J. van der Veen (1993) 'Activering van beleidsnetwerken rond milieuproblemen', in: J.F.M. Koppenjan, J.A. de Bruijn & W.J.M. Kickert (red.) *Netwerkmanagement in het openbaar bestuur. Over de mogelijkheden van overheidssturing in beleidsnetwerken,* VUGA Uitgeverij, Den Haag, pp. 131-146.

Driessen, P.P.J., P. Glasbergen, P.P.P. Huigen & F. Hijmans van den Bergh (1995) *Vernieuwing van het landelijk gebied; een verkenning van strategieën voor een gebiedsgerichte aanpak,* VUGA Uitgeverij, Den Haag.

Faulks, K. (1999) *Political Sociology. A Critical Introduction,* Edinburgh University Press, Edinburgh.

Giddens, A. (1994) *Beyond Left and Right,* Polity Press, Cambridge.

Glasbergen, P. (1989) *Beleidsnetwerken rond milieuproblemen,* VUGA Uitgeverij, Den Haag.

Godfroij, A.J.A. (1981) *Netwerken van organisaties: strategieën, spelen, structuren,* VUGA Uitgeverij, Den Haag.

Godfroij, A.J.A. & N.J.M. Nelissen (1993) *Verschuivingen in de besturing van de samenleving,* Coutinho, Bussum.

Goverde, H. & J. van Tatenhove (2000) 'Power and Policy Networks', in: H. Goverde, Ph.G. Cerny, M. Haugaard & H.H. Lentner (eds) *Power in Contemporary Politics. Theories, Practices, Globalizations,* Sage, London, pp. 96-111.

Kickert, W.J.M., E-H. Klijn & J.F.M. Koppenjan (eds) (1997) *Managing Complex Networks. Strategies for the Public Sector,* Sage Publications, London.

Klijn, E-H. (1996) *Regels en sturing in netwerken, De invloed van netwerkregels op de herstructurering van naoorlogse wijken,* Eburon, Delft.

Klijn, E-H. (1997) 'Policy Networks: An Overview', in: W.J.M. Kickert, E-H Klijn & J.F.M. Koppenjan (eds) *Managing Complex Networks. Strategies for the Public Sector,* Sage Publications, London, pp. 14-34.

Klijn, E.H. & G.R. Teisman (1997) 'Strategies and Games in Networks', in: W.J.M. Kickert, E-H Klijn & J.F.M. Koppenjan (eds) *Managing Complex Networks. Strategies for the Public Sector,* Sage Publications, London, , pp. 98-118.

Marin, B. & R. Mayntz (eds) (1991) *Policy Networks. Emprical Evidence and Theoretical Considerations,* Campus Verlag/Westview Press, Frankfurt am Main, Boulder, Colorado.

Nelissen, N.J.M. (1993) 'Over 'net' werken in netwerken', in: J.F.M. Koppenjan, J.A. de Bruijn & W.J.M. Kickert (red.) *Netwerkmanagement in het openbaar bestuur. Over de mogelijkheden van overheidssturing in beleidsnetwerken,* VUGA Uitgeverij, Den Haag, pp. 169-175.

Nelissen, N.J.M. (1994) 'De terugtredende overheid: wenkend of beknellend perspectief?', in: *Preadvies Vereniging voor Milieurecht,* Zwolle, pp. 1-66.

Rhodes, R.A.W. & D. Marsh (1992) 'New Directions in the Study of Policy Networks', *European Journal of Political Research,* 21 (1-2), p. 181-205.

Teisman, G.R. (1992) *Complexe besluitvorming. Een pluricentrisch perspektief op besluitvorming over ruimtelijke investeringen,* VUGA Uitgeverij, Den Haag.

Termeer, C.J.A.M. (1993) *Dynamiek en inertie rondom mestbeleid; een studie naar veranderingsprocessen in het varkenshouderijnetwerk,* VUGA Uitgeverij, Den Haag.
Voogd, A.A. (1990) *Managen in een meervoudige context; naar een methode voor het ontwikkelen van sociaal-cognitieve configuraties,* Eburon, Delft.
Vos, G. et al. (1994) *Sturing op maat. Een andere benadering van milieuproblemen in de land- en tuinbouw,* Ministerie van LNV, Den Haag.

Chapter 3

STRATEGIES IN ENVIRONMENTAL POLICY

A historical institutional perspective

Jan P.M. van Tatenhove and Henri J.M. Goverde

3.1 Introduction

The institutionalisation of environmental policy in the Netherlands will be discussed in this chapter. Institutionalisation is a process that constructs and preserves day-to-day activities and interactions within a context of societal and political change. Institutionalisation is one of the central concepts used in the social and political sciences to study the ongoing process of transformation in different polities. "Social, political and economic institutions have become larger, considerably more complex and resourceful, and 'prima facie' more important to collective life" (March & Olson, 1984, p. 734). This 'new' institutionalism emphasises the relative autonomy of political institutions, the possibilities for inefficiency in history, and the importance of symbolic action to an understanding of politics (March & Olson, 1984). Keman (in: Steunenberg et al., 1996) distinguishes between three approaches within neo-institutionalism: the historical-traditional approach, a cultural-organisational approach, and the (political-economic) rational choice approach. In order to make a choice we will follow the argument put forward by Bulmer (1998).

Two aspects are particularly distinctive to new institutionalism according to Bulmer. One is a wider interpretation of what constitutes institutions. While there is a shift away from formal constitutional-legal approaches to governance, the less formalised arenas of politics are still taken into account. This is why the new institutionalism has become sensitive to the findings set out in 'policy community' and 'policy network' literature (Bulmer, 1998, p. 369). The second distinction is a concern with

P.P.J. Driessen and P. Glasbergen (eds.), Greening Society, 47–63.

the beliefs, paradigms, codes, cultures and knowledge embedded within the institutions. In this sense the new institutionalism is an attempt to correct overly rationalist and functionalist interpretations in political science. In his critique on rationalist approaches in institutionalist theory, Bulmer introduces historical institutionalism. Historical institutionalist analysis includes the normative and cultural dimensions, both of which go beyond rationalist calculations. Historical institutionalism is subsequently analysed by identifying four dimensions: systemic change, governance structures, policy evolution and the role of values and norms.

In this chapter we develop a framework of institutionalisation which is comparable with Bulmer's historical institutionalism, in which there is a duality of systemic change and strategic conduct. In general, institutionalisation is a process of production and reproduction of strategies, including a range of policy arrangements, as a result of the duality between changing relations between state, market and civil society on the one hand (systemic change) and the norms, arguments and values actors use in their interaction on the other (strategic conduct). Strategies are perceived here as reconstructions at the national level of the power relations between different ministries dealing with the environment, and the way public actors operate in shifting interdependencies between state, civil society and market (Van Tatenhove, 1993). In the next sections we will discuss the following strategies:

- The demarcation and ordering strategy.
- The break-in-strategy.
- The societalisation strategy.
- The regionalisation strategy.
- The broadening strategy.
- The flexibility strategy.

The 1970s and 1980s strategies – i.e. the demarcation and ordering strategy, the break-in strategy, the societalisation strategy and the regionalisation strategy – are all expressions of the ongoing struggle to give environmental issues a position in other domains of policy. This continuous struggle has to be understood in the context of changing interrelations between state, market and civil society. During the 1990s there was a renewal of policy arrangements, expressed in the broadening and flexibility strategy. In general terms, each strategy affects the dimensions, coalitions, rules, resources and discourses in environmental affairs in a specific way, resulting in different patterns of change and stability in environmental policy. The content of these different strategies demonstrates a wide variety of environmental policy arrangements. A policy arrangement refers to the temporary stabilisation of the organisation and substance of a policy domain. As such it is the outcome of interactions and interdependencies among actors in the context of processes of structural change, that is political modernisation (Van Tatenhove et al., 2000). In general, policy arrangements are analysed on the basis of four dimensions: coalitions; rules; resources and discourses.

3.2 The demarcation and ordering strategy

The end of the 1960s is a valid starting point for describing the institutionalisation of environmental policy. At that time there was considerable concern about the environment among scientists, and public debates about the seriousness of environmental problems were starting to appear. These debates were followed by a growing political and general interest in environmental problems. Characteristic of the institutionalisation of Dutch environmental policy in the 1970s and 1980s was the 'demarcation and ordering strategy'. To understand this strategy we will first discuss the initial steps in the process of political modernisation during the heyday of 'early' and (political) 'anti' modernisation (Van Tatenhove, 1999). On the one hand there was a great optimism to solve (urgent) environmental problems with technical measures and legislation. End-of-pipe technologies were used to clean up the most threatening of the environmental problems such as water pollution, air pollution, waste and noise. On the other hand, the scientific and societal environmental concern resulted in the foundation of ministries or administrative bodies to deal with environmental problems. In the Netherlands, the Ministry of Public Health and Environmental Hygiene was established in 1971. In the same period, administrative bodies were institutionalised in other Western countries, such as the Environmental Protection Agency in the US (1970), the German Interior Ministry (1969), the Ministry for the Protection of Nature and the Environment in France (1971) and the Department of the Environment in the UK (1970). "A common feature of substantive environmental policy in these Western countries in the 1970s was the use of traditional administrative regulatory strategies. Policy-making relied to a large extent on direct regulation: general standards were set at the national level and formed the basis of licenses for individual plants or activities" (Van Tatenhove, 1999). Regulation by legal rule was the norm. Examples of this kind of legislation are the 1970 Clean Air Act in the United States, the 1974 Federal Emission Control Act in the Federal Republic of Germany, the 1974 Control of Pollution Act in the United Kingdom, the 1961 Air Pollution Law and the 1964 Water Law in France, and the 1969 Surface Water Pollution Act and the 1970 Air Pollution Act in the Netherlands (Weale, 1992).

Regulation by legal rule and regulatory strategies in the Netherlands resulted in the Urgency Memorandum (1972) and an impressive body of sectoral legislation. Characteristic of Dutch environmental policy is what Verkruisen (1978) referred to as a double problem. On the one hand there was a kind of urgency to solve environmental problems, while on the other hand there was no shared opinion about the position of environmental policy in relation to other domains of policy. This resulted in a fragmented institutionalisation of environmental policy; the competency for the design and implementation of environmental policy was divided among a number of ministries and administrative bodies, all dealing with different aspects of the environment (Van Tatenhove, 1993). First of all, this fragmentation complicated the integration and implementation of policy since every single citizen and entrepreneur applying for a permit had to deal with totally different governmental bodies, all looking separately at a single aspect. Secondly, the fragmented institutionalisation caused the problem of cross-media transfer because the policy was only intended to protect one particular environmental medium, such as air, thereby neglecting the fact that regulating air pollution could easily create landfill or water pollution problems (Weale, 1992). Thirdly, given that separate legislation and regulations were developed for each compartment, plus the inadequate coordination of licensing procedures and enforcement, there is also procedural fragmentation. This gives rise to the question of legal and administrative coordination.

As a result of this process of fragmented institutionalisation the Ministry of Housing, Spatial Planning and the Environment pursued a demarcation and ordering strategy. The aim of this strategy was to establish its position among the other administrative bodies that dealt with the environment. In the 1980s the new minister for the environment, Winsemius, set out this strategy in greater detail. With the introduction of 'the environmental management philosophy' a process of internal institutionalisation was set in motion. Characteristic of this environmental management philosophy are three elements. Firstly, the introduction of management models such as the 'regulation chain', the 'integration column' and the 'policy life cycle'. Secondly, the introduction of the so-called two-track policy, focusing both on the sources and the effects of pollution. Thirdly, this philosophy involved the introduction of new objectives such as 'sustainable development' and 'internalisation'. This discursive innovation in environmental policy brought other coalitions into view which gave shape and form to environmental policy on the basis of new regulations.

With regard to the management models, the policy life cycle first of all demonstrates that the development of environmental policy is a step-by-step process. That is from recognition, through the formulation of policy, to solving the problem and control. Furthermore the degree of conflict on the point at issue as well as the political importance of the issue differ in each stage. The regulation chain shows that various stages can be distinguished in the juridical management of environmental issues (legislation and regulations, the granting of permits, implementation and enforcement). The integration column illustrates which tasks face environmental policy. A distinction is also made between internal and external integration. Internal integration refers to the achievement of cohesion and consistency between the different components of environmental policy as pursued by the ministries involved. External integration concerns the performance of cohesion and consistency between environmental policy and other elements of government policy (Ministry of Housing, Spatial Planning and the Environment, 1983; Winsemius, 1986, pp. 15-19, 63, 80). These models are concerned not only with structuring the actions of politicians and public servants, but also serve to detect the weak and strong points of environmental policy in terms of organisation and content.

The second pillar of policy philosophy is the two-track policy: source-oriented and effect-oriented policy. One specific effect of the source-oriented track was pivotal in the 1970s (the granting of individual permits, preventing unnecessary pollution, prevention at source, etc.). The effect-oriented track was worked out in detail in the 1980s. The point of departure was to ensure that there would be no adverse effects in both the short and long term for human beings, plants, animals and the ecosystems. This brings the quality of the physical environment specifically into the picture in the form of environmental quality requirements and risk control. The source-oriented policy of the 1980s focused on preventing unacceptable risks. The point of departure here was to protect the quality of the physical environment by making use of all the available opportunities to minimise emissions and other forms of environmental degradation at source.

It was suggested in the first National Environmental Policy Plan (NEPP) that the environmental problems were much more serious than had been estimated. The strategy would therefore need to be adjusted; a question of breaking open and changing the material cycles and combating the mechanisms for shifting responsibility. The capacity of the environment regarding load is seen in terms of a unit of stock, in the same way as we see the stock of raw materials. The flow of materials between these stocks must be controlled if we are to maintain the environment's capacity. This control can be achieved through separating the economic and ecological material flows by preventing loss from leakage in the economic flow. As long as this has not been achieved, the external effects of the economic flow of materials must be controlled (Ministry of Housing, Spatial Planning and the Environment, 1989, p. 9). Other arguments are also put forward to introduce controlled feedback mechanisms in order to prevent responsibility from being shifted elsewhere (to other levels of scale, other societal groups or other generations). This leads to a distinction between feedback at source (integral chain management, energy extensification and the promotion of quality) and feedback by way of risk management. Source-oriented measures are divided into emission-based measures, volume-based measures and structural measures. While the NEPP gives preference to structural, source-oriented measures, effect-oriented measures are still required in order to tackle our legacies from the past. The essence of source-oriented policy is the direct approach on (potential) sources of pollution.

The third pillar of the contextual restructuring of environmental policy consists of two new points of departure: the internalisation of environmental responsibilities and sustainability. The concept of 'internalisation', introduced to improve the enforcement of environmental legislation in the industrial sector, soon evolved into a discourse within Dutch environmental policy to make the polluters aware of their environmental responsibilities. On the one hand internalisation is concerned with the necessity to boost the polluters' willingness to contribute towards solving environmental problems by appealing to their sense of 'social responsibility'. On the other hand, it gives serious attention to the needs and wishes of groups affected by environmental policies, and taking them into account in the design of measures (Liefferink & Mol, 1996). The concept of internalisation is elaborated in 'target group policy' (see also section 3.4). Internalisation implies the integration of environmental aspects in the actions of the target groups.

The main concept of the Brundtland Commission, that of 'sustainable development', was taken over as the main point of departure for the first National Environmental Policy Plan. The principal goal of environmental policy is 'to retain the environment's capacity in order to achieve a sustainable development' (Ministry of Housing, Spatial Planning and the Environment, 1989, p. 92). Sustainable development meets "the needs of the present generation without endangering the options available to future generations for them to meet their needs too" (Ministry of Housing, Spatial Planning and the Environment, 1989, p. 42).

The formulation of such broad, comprehensive goals, such as sustainability and internalisation is directional for the actions taken by those responsible for implementing policy, and – as we shall see in the break-in strategy – simultaneously expands the domain of the Ministry of Housing, Spatial Planning and the Environment.

3.3 The break-in strategy

In the 1980s the Ministry of Housing, Spatial Planning and the Environment tried to overcome the problems caused by the fragmented institutionalisation, especially the struggle that had emerged among various ministries for the responsibility for environmental policy. The nature and content of this internal institutionalisation was subject to the emphatic influence of external social and cultural factors such as society's awareness of the environment (Nelissen, 1987), the stand taken by the environmentalist movement, and the analyses of the environmental problems by political parties, the environmentalist movement, scientists and advisory bodies. Despite several crises the efforts of both the environmentalist movement and Dutch Parliament ensured that the environmental issues were kept on the political and societal agenda. The environmentalist movement was an important resource for the Ministry of the Environment in the interdepartmental struggles for power. The relationship between the environmentalist movement and the ministry in the 1970s was still a relatively hostile one. According to Cramer (1989, p. 122), however "(...) the 1980s saw an increase in their understanding of one another (...); the environmentalist movement had to admit that the ministry was prepared to do its absolute best for the environment. This aroused the environmentalist movement to support the Ministry of Housing, Spatial Planning and the Environment in the ministerial battle".

Another resource used by the Ministry of the Environment in the interdepartmental battle for policy were the national advisory bodies. The 'Centrale Raad voor de Milieuhygiëne' *(Environmental Protection Advisory Council)* was established in 1974 under pressure from the Parliament and the environmental organisations. Ms Vorrink, who was a minister at the time, referred to this Council as the 'environmental protection conscience' of the Netherlands. Its members were the representatives of consumer and environmental organisations, employers' organisations and trade unions, utility companies, the Industry Board for Agriculture, local authorities, water control authorities and independent experts. Despite the frequent lack of unanimity within the Council, it was still a significant source of information for government and parliament regarding the arguments put forward by non-governmental organisations. Some years ago the Council was transformed into the 'VROM-raad', or *Ministry of Housing, Spatial Planning and the Environment Council,* and now combines advice on environmental management with advice on public housing and spatial planning.

By the end of the 1980s the environmentalist movement and advisory bodies had become highly professional in terms of organisation and content. Hence their contribution in the policy-making process started to be taken more seriously in the established political system. The policy makers within the Ministry of Housing, Spatial Planning and the Environment used the external social and political support as a catalyst to impede its 'break-in' strategy in this interdepartmental struggle in order to secure the acceptance of environmental perspectives and objectives in other fields of policy. A good example of the 'break-in' strategy was the completion of the first National Environmental Policy Plan in 1989. This Plan was the result of negotiations between several ministries, and although it bears the characteristics of a compromise the binding element of the different views of the participating ministries was the concept of sustainable development as originally introduced in the Brundtland report (WCED, 1987).

However, the negotiations on the draft versions of the first NEPP between the Ministry of the Environment, the Ministry of Agriculture, the Ministry of Transport and Public Works and the Ministry of Economic Affairs were very laborious. There were two distinct phases in these negotiations. Emphasis in the first phase was on analysing environmental problems and formulating the points of departure. Given that the Cabinet had already adopted the point of departure of sustainable development as government policy in October 1987, the other ministries had little problem in subscribing to the general analysis of the environmental problems. Nevertheless, problems arose when interpreting the general points of departure into specific measures in the various fields of policy. In this second phase, the Ministry of Agriculture, the Ministry of Transport and Public Works and the Ministry of Economic Affairs were apparently too nervous about being made responsible for environmental measures that might be too strict. The Ministry of Housing, Spatial Planning and the Environment therefore chose to pursue an offensive divide and rule strategy. By making the environmental departments of the Ministry of Agriculture (nature conservation) and the Ministry of Economic Affairs (energy saving) partners in environmental policy, the Ministry of Housing, Spatial Planning and the Environment attempted to establish a link with these predominantly economically-oriented ministries. This strategy was only a partial success; the tension between economic matters and the environment remained. The other ministries apparently had no wish to work on the lines of the Ministry of Housing, Spatial Planning and the Environment as yet. The benefit of the strategy was that the other fields of policy were broadened to include the environmental aspect (Van Tatenhove, 1993, pp. 43-44).

3.4 The societalisation strategy

Again, this strategy is an answer to the fragmentation of environmental policy among different ministries. The societalisation strategy is directed towards a variety of target groups located in policy fields such as agriculture, industrial affairs and traffic, particularly lorries and private cars. By contacting these groups directly an attempt is made to bring about an 'internalisation' of environmental perspectives among them. Both 'internalisation' and target group policy (see: Nelissen et al., 1988; Van Vliet, 1992; LeBlansch, 1996) are an answer to the shortcomings within environmental policy and a reaction to external developments. Internally, the enforcement and implementation structures were the bottlenecks of environmental policy, while externally, the institutionalisation of environmental politics was radically affected by the changing relationships between state, market, and civil society, resulting in new ways of governance and participation. The target groups in Dutch environmental policy are defined as more or less homogeneous groups of polluters, such as the agricultural sector, the traffic and transport sector, industry and refineries, gas and electricity suppliers. Target group policy has evolved from 'target group as the object of management' to 'target group as interlocutor and negotiating partner', and has now expanded into a general policy philosophy. The essence of this philosophy is that 'a pollution prevention and source-oriented approach towards the environmental problems which is in line with the aims to achieve sustainable development, is dependent on the cooperation and accountability of the economic sectors concerned' (Van Vliet, 1992, p. 100).

Pivotal for the societalisation strategy is that the impeding policy structures of other ministries can be circumnavigated. To be more straightforward: closer contact with the traditional target groups of the Ministry of Agriculture (farmers and their organisations) or the Ministry of Economic Affairs (industry, energy sector, etc.) may prevent those ministries from translating, amending and even underexposing the environmental 'message' to make it fit the productivist and economic orientation of the economic sector in its broadest sense. Hence, both the target group and the internalisation discourse are concrete elaborations of the societalisation strategy. Although the final goal is to prevent environmental pollution, this strategy also serves to strengthen the ministry's coordinating tasks by ensuring that environmental goals are accepted in the policy fields to which these target groups belong.

3.5 The regionalisation strategy

This strategy is also a reaction to fragmented institutionalisation. However, it does not focus on individual ministries or target groups but on the regionalisation of policy. It is becoming increasingly clear that generic environmental standards are important, but that the success of implementing and enforcing those standards depends on the specific regional circumstances and the relations between the various parties within a certain region. The 'regionalisation strategy' consists of introducing region-oriented environmental policy and involves shifting the administrative responsibilities from the national to regional and local levels. Region-oriented policy is seen as a framework for policy integration by combining environmental and spatial planning at the regional level. 'Integrated region-oriented policy' reflects the changed relations between state and society. Although the initiative is in the hands of public actors, the policy measures themselves are developed in association with private actors. The aspects of cooperation, negotiation and deliberation play a major role in this process. The 'integrated region-oriented approach' can also be seen as a way to overcome the problems of internal and external integration. One of the main goals is to integrate different policy fields in regional development. Shifting the administrative responsibilities by way of regionalisation and decentralisation passes on the predicaments of national environmental policy to local authorities.

Besides regional projects with a predominantly regulatory nature, projects characterised by a more or less integrative approach and reflexive forms of governance have also been developed. The first category comprises mainly projects that are oriented towards a specific policy domain such as groundwater protection areas, the preservation of nature and wetlands, where the emphasis is on relatively strict rules for environmental protection. The second category, for example, consists of area-based integrated 'ROM-projects' *(Spatial Planning and Environmental projects)* and a group of somewhat heterogeneous integrated region-oriented projects such as 'City and the Environment' and ROM like projects including Valuable Man-made Landscapes (VML) and Rural Area Pilot Plans (RAPP) (see also chapters 6 and 7). The so-called ROM-projects were initiated and implemented at the beginning of the 1990s. They represent a particular group of ten areas in which there is enormous pressure on the environment and spatial qualities, or pressure is anticipated in the near future. In these projects, governmental agencies and private parties in the ROM field are invited to draw up a working plan to solve acute environmental problems, to improve spatial planning characteristics and to implement a package of concrete measures and promising socioeconomic projects. Specific regional circumstances are taken into account in this latter category by differentiating norms.

All in all, the regionalisation strategy is one of the many reactions to the internally fragmented institutionalisation of environmental policy over the period 1970-1985. At the same time, the changed relations between the government, the market and civil society are taken advantage of, especially in the second half of the 1980s. This resulted in a considerable change in the semblance of environmental policy. Area-oriented environmental policy (including the decentralisation of administrative responsibilities to lower administrative levels) becomes the third track in environmental policy alongside target groups and specific themes. Another development that can be ascertained in area-oriented environmental policy is the sectoral approach taken to achieve a method of approach that focuses on external integration.

3.6 The renewal of policy arrangements: broadening and flexibility

The 1990s was a period in which the already institutionalised policy arrangements were consolidated and environmental policy was renewed. Apart from that, the debate on integration and the organisational embedment thereof was influenced by new forms of administration (channelling the market, providing local authorities with scope for policy-making) and new forms of management (network management and self-regulation), and also by an increase in the number of structural societal processes (individualisation, europeanisation and globalisation). There was a certain amount of tension between stability and change in Dutch environmental policy. Further development of the target group and integrated region-oriented policies implied the consolidation of environmental policy, while at the same time basically altering its principles of governance. The latter is best summarised in the concepts of broadening and flexibility. Broadening here refers to the wish to achieve integral green planning, in which sustainability is an integrated element of other sectoral activities and policies. Flexibility implies the development of interactive and reflexive types of policy making. Increasingly, the formulation and implementation of environmental policy is the outcome of interactions and negotiations between public and private actors.

Four conferences organised between 1990 and 1996 gave momentum to the broadening and flexibility of environmental policy content. Representatives from the Ministry of Housing, Spatial Planning and the Environment, the Union of Water Boards, the Inter-provincial Consultative Body and the Association of Dutch Municipalities attended these conferences. The main conclusions were that environmental policy is the concern of all parties involved and not a state monopoly, that there must be a change from government to governance, and that – subject to certain conditions being met – it is acceptable to deviate from environmental norms. This new stance was set out in the third and fourth National Environmental Policy Plan (Ministry of Housing, Spatial Planning and the Environment, 1998, 2001). While implementation and the further development of target group policy was emphasised in NEPP2, in NEPP3 emphasis was also given to stimulating sustainable patterns of production and consumption, differentiating environmental norms for different regions, and external integration with physical planning and water management (Janssens & Van Tatenhove, 2000).

The broadening and flexibility of environmental policy is elaborated in both the discourses on integration and governance (Janssens & Van Tatenhove, 2000). On the one hand the integration discourse recently forced the domains of environmental policy, spatial planning and water management to increase the coordination and integration of their planning activities. On the other hand the discourse on governance has resulted in a shift from state-dominated policy-making to policy-making processes in which market relationships and the participation of actors from civil society has become more prevalent. Not only is the ideal of a clean environment of importance in the aim towards a liveable environment, but the intention is also to increase the level of support for, and the level of involvement in environmental policy.

The aim to broaden environmental policy generally brings about a shift from internal to external integration. External integration (the integration of environmental aspects into a variety of other components of government policy and economic activities) should be reinforced. To this end, a policy document on the Environment and the Economy (Ministry of Housing, Spatial Planning and the Environment et al., 1997) was drawn up at central government level. At the provincial level this is reflected in the aim to draw up an integrated plan, integrating the plans of different policy domains like environmental policy, spatial planning and water management, and to implement several region-oriented projects. At the level of the local authorities, the 'City and the Environment' projects are worthy of note (see Chapter 7). This trend reflects an on-going societalisation and regionalisation of environmental policy.

3.7 Comments on the broadening and flexibility of environmental policy

A few comments must be made regarding the wish to broaden environmental policy and make it more flexible. The first relates to the aim of broadening environmental policy. This wish is expressed in a context within which the structure of environmental policy – in terms of both content and organisation – is still somewhat fragmented. This level of fragmentation is both acknowledged and questioned by the Advisory Council on Government Policy (WRR, 1998, p. 114) which states that "parties in several overlapping fields of policy have expressed their views regarding the aspects that should play a role in the policy-making process". This has resulted in the circulation of different definitions of integrality. According to the Advisory Council on Government Policy, the aim towards a policy on the natural environment (integration of the environment and spatial development) reinforces the sectoral character of policy pursued by the Ministry of Housing, Spatial Planning and the Environment. The quality of the natural environment is seen increasingly as a specific interest which must be embedded in the processes of weighing up the various interests (WRR, 1998, p.114). Similar criticism was expressed by Bouwer (1997) who said that the drawback of the course that had already been taken to broaden policy was that external integration is elaborated upon and detailed, while internal integration is not yet even discussed. Environmental goals and points of departure are thus ignored in an ongoing battle with other economic and societal interests.

A comment can also be made in connection with the aim to make environmental policy more flexible. This flexibilisation in environmental policy is expressed in new principles of governance: 'customised management', interactive policy and differentiated standards. Characteristic of the flexibility in environmental policy is the changed role of the public actors. Policy is increasingly formulated and implemented in dialogue with private actors. This societalisation of environmental policy is reflected in the forming of new arrangements within which the actors – in interactive and communicative forms of work – search (collectively if at all possible) for new coalitions, policy concepts, rules and resources with which to tackle environmental issues. However, these new arrangements are not a substitute for the traditional policy arrangements. Legislation and planning regimes are the main weapons for the government to use as a last resort. Although the government now adopts a more reticent attitude, its position remains to be a special one. It can always be requested to act as the arbitrator by societal actors, for instance within the framework of guaranteeing fundamental rights and legal certainty (see also Chapter 2).

The internal and external institutionalisation of environmental policy in the 1990s on the lines of broadening and flexibility led to wide diversity in terms of policy arrangements. Characteristic of these arrangements is that management is in the hands of many different actors. Not only are new environmental goals formulated in negotiations between the various policy coalitions in integrated area-oriented projects in rural areas, for instance, such as ROM projects, VMLs (Valuable Man-made Landscapes) and SGPs (Strategic Green Projects), but there is also the opportunity for interest groups to change the 'rules of the game' and to effectively mobilise resources. It would seem that the Ministry of Housing, Spatial Planning and the Environment is consciously leaving the management of these policy arrangements to others.

One of the advantages of the 'regionalisation' of environmental policy in particular is that it then becomes possible to boost support for environmental measures. This leads to our third comment: that this method of approach could disguise the danger that the definition of environmental problems might be limited to such themes as disturbance and habitat fragmentation, themes that can be resolved on the regional level in consultation with the actors concerned. Other major environmental themes such as climate change, energy and raw materials, are therefore in danger of becoming the balancing items of environmental policy which must then be negotiated against (spatial) economic and infrastructural interests until a solution is found. It is obvious that the regionalisation of environmental policy is not the only way to achieve environmental goals. After all, typical of new policy arrangements is that coalitions of actors are not restricted to one level alone, nor that the arrangements are managed centrally. In addition to the regionalisation of environmental policy there is also a 'transnationalisation' or even a 'globalisation' of environmental policy. The definition of environmental problems therefore takes place on different levels. Constantly changing actor coalitions are involved: not only national states, multinationals and global NGOs, but also local NGOs, citizens, small and medium-sized businesses, etc. Potential options formulated in international forums must then be translated and implemented in national institutions, and thus in a national context of traditional, but above all new policy arrangements.

3.8 Conclusions

This chapter has defined and analysed the institutionalisation of environmental policy since the 1970s on the basis of six strategies. These strategies represent the paradigm shift that has taken place in environmental policy over the past thirty years. In brief, this paradigm shift (see Chapter 1) implies that under the influence of changing relations between the state, the market and civil society, changes have taken place in both the strategic way of thinking in terms of the content and organisation of environmental policy, and in the procedure chosen. Shifts can be observed in the competence fields of government, trade and industry and non-profit organisations. Nelissen (1992) and Goverde (1993) discern etatistic, market, corporatist and citizen patterns. These patterns of shift reflect a change in the powers of parties in the fields of the state, the market and civil society. If we take a closer look at environmental policy we see that these patterns of shift reflect the strategies set out in this chapter and the range of new policy arrangements that have been formed as a result of this trend.

In the 1970s the institutionalisation of environmental policy was mainly a state responsibility. Assisted by the demarcation and ordering strategy and the break-in strategy, an attempt was made to give the environmental interest a place in other fields of policy and in other ministries. Etatistic policy arrangements were formed, within which central government formulated urgencies and – helped by sectoral environmental protection legislation and plans – attempted to improve the quality of the physical environment. However, as a result of an interdepartmental battle for the demarcation of environmental policy, the ordering or demarcation strategy and the break-in strategy were evidently only a partial success. The result was organisational fragmentation: environmental policy was given a place in a multitude of departmental agencies and institutions.

Internal and external factors brought about a change in the institutionalisation of environmental policy in the early-1980s. One of the main internal factors was the establishment of the Ministry of Housing, Spatial Planning and the Environment, making it easier to establish the position of environmental policy in comparison with other fields of policy. Nevertheless, the ministry was at the same time faced with the problems of implementing and enforcing environmental policy, a legacy from the 1970s when the implementation and enforcement of certain parts of environmental policy were delegated to other ministries. A major external factor was the erosion of the concept of the state's makeability of society. This led to new strategies in which parts of the market and civil society became involved in realising the implementation of environmental policy. In concrete terms, this was expressed in the societalisation and regionalisation strategies. Both strategies implied a renewal of environmental policy in terms of content and organisation. Regarding content, the relationships between the separate environmental compartments (water, soil and air) were stressed by way of themes and area-oriented policy. Moreover, groups of polluters were discerned (target group policy) and environmental problems were formulated on several levels of scale (local, regional, fluvial, continental and global). This broadening of environmental policy content was subsequently translated into a new organisational direction: consultation with target groups (agreements); 'leaving policy up to others' (local authorities and private parties); differentiation in standards; demarcated self-regulation; facilitation, etc. In other words: different forms of liberal and corporatist policy arrangements emerged in which central government redefined the rules and resources of environmental policy in consultation with societal interest groups and market parties.

The changing relations in the 1990s between state, market and civil society had a strong influence on the institutionalisation of environmental policy. The broadening and flexibilisation strategy of environmental policy led to the process of environmental policy institutionalisation no longer being an internal government matter. This is evident, for instance, from how environmental problems are defined. The definition of environmental problems and the discursive innovation of environmental policy are no longer the sole domain of the state, but the result of negotiations among a variety of actors at different levels of scale (ranging from local to global). Each actor constructs environmental problems in a certain way, depending on the societal interests one represents. When searching for solutions the environmental interest is thus a constant part of the battle for power between different interests: the economy, space, infrastructure, etc. Given relationships are not very stable in such reflexive policy arrangements. For instance, the target group policy reflects the characteristics of a neo-corporatist policy arrangement, in which the elites of government, trade and industry formulate common policy for the entire branch of industry in consultation and in accordance with set rules. However, trade and industry are no longer 'condemned' to the national government but can also enter into coalitions with other stakeholders, e.g. environmental organisations. For the government this means that there are no longer any obvious, 'natural' policy addressees. The search for solutions to environmental problems will therefore result in a continuing diversification of policy arrangements. The actual predictability of policy thus diminishes, yet conversely, the desired flexibility increases. This necessitates a rethink as to the 'director's role' in environmental policy. Who is able to, and indeed must take on this role? In addition to this question of organisational institutionalisation, the leading principles on which environmental policy leans in terms of content will probably – under the influence of the multitude of arrangements – be brought up for discussion yet again. Aided by the historical institutional perspective used in this chapter, it is quite possible to describe and analyse this dynamism of institutionalisation processes on the basis of strategies and developing policy arrangements.

References

Bouwer, K. (1997) *Van milieubeleid naar omgevingsbeleid?,* Katholieke Universiteit Nijmegen, Nijmegen.

Bulmer, S.J. (1998) 'New Institutionalism and the Governance of the Single European Market', *Journal of European Public Policy,* (5) 3, pp. 365-386.

Cramer, J. (1989) *De Groene Golf. Geschiedenis en toekomst van de Nederlandse milieubeweging,* Jan van Arkel, Utrecht.

Goverde, H.J.M. (1993) 'Verschuivingen in het milieubeleid; van milieuhygiëne naar omgevingsmanagement', in: A.J.A. Godfroij en N.J.M. Nelissen (red.) *Verschuivingen in de besturing van de samenleving,* Coutinho, Bussum, pp. 49-88.

Janssens, J. & J. van Tatenhove (2000) 'Green Planning: From Sectoral to Integrative Planning Arrangements?', in: J. van Tatenhove, B. Arts & Pieter Leroy (eds) *Political Modernisation and the Environment. The Renewal of Environmental Policy Arrangements,* Kluwer Academic Publishers, Dordrecht, pp. 145-174.

Le Blansch, C.G.L. (1996) *Milieuzorg in bedrijven. Overheidssturing in het perspectief van de verinnerlijkingsbeleidslijn,* Thesis Publishers, Amsterdam.

Liefferink, J.D. & A. Mol (1996) *Voluntary Agreements as a Form of Deregulation? The Dutch Experience.* Paper prepared for the Working Group on Environmental Studies, European University Florence, 9-11 May.

March, J.G. & J.P. Olson (1984) 'The New Institutionalism; Organizational Factors in Political Life', *The American Political Science Review,* 78 (3), pp. 734-749.

Ministry of Housing, Spatial Planning and the Environment (Ministerie van VROM) (1983) *Plan Integratie Milieubeleid,* TK 18010, Den Haag.

Ministry of Housing, Spatial Planning and the Environment (Ministerie van VROM) (1984) *Meer dan de som der delen,* TK 18292, Den Haag.

Ministry of Housing, Spatial Planning and the Environment (Ministerie van VROM) (1989) *Nationaal milieubeleidsplan. Kiezen of verliezen,* TK 21137, Den Haag.

Ministry of Housing, Spatial Planning and the Environment (Ministerie van VROM) (1997) *Nota milieu en economie. Op weg naar een duurzame economie,* TK 25405, Den Haag.

Ministry of Housing, Spatial Planning and the Environment (Ministerie van VROM) (1998) *Nationaal Milieubeleidsplan 3,* Den Haag.

Ministry of Housing, Spatial Planning and the Environment (Ministerie van VROM) (2001) *Nationaal Milieubeleidsplan 4. Een wereld en een wil,* Den Haag.

Nelissen, N.J.M. (1987) *De Nederlanders en hun milieu. Een onderzoek naar het milieubesef en het milieugedrag van vroeger en nu,* Kerckebosch, Zeist.

Nelissen, N., H. van Boxtel, E. Houben & M. Lemmen (1988) *Het milieu: vertrouw, maar weet wel wie je vertrouwt. Een onderzoek naar verinnerlijking en verinnerlijkingsbeleid op het gebied van milieu,* Kerckebosch, Zeist.

Nelissen, N.J.M. (1992) *Besturen binnen verschuivende grenzen,* Kerckebosch, Zeist.

Steunenberg, B., J. de Vries & J.L. Soeters (1996) 'Het neo-institutionalisme in de bestuurskunde', *Bestuurskunde,* (5) 5, pp.212-216.

Van Tatenhove, J. (1993) *Milieubeleid onder dak? Beleidsvoeringsprocessen in het Nederlandse milieubeleid in de periode 1979-1990, nader uitgewerkt voor de Gelderse Vallei,* Pudoc, Wageningen.

Van Tatenhove, J. (1999) 'Political Modernisation and the Institutionalisation of Environmental Policy', in: M. Wissenburg, G. Orhan & U. Collier (eds.) *European Discourses on Environmental Policy,* Ashgate, Aldershot, pp. 59-78.

Van Tatenhove, J., B. Arts & P. Leroy (eds.) (2000) *Political Modernisation and the Environment. The Renewal of Environmental Policy Arrangements,* Kluwer Academic Publishers, Dordrecht.

Van Vliet, L.M. (1992) *Communicatieve besturing van het milieuhandelen van ondernemingen: mogelijkheden en beperkingen,* Rotterdam.

Verkruisen, W.G. (1977) De plaats van het milieubeleid in het regeringsbeleid, *Milieu & Recht,* nr. 4, pp. 8-14.

WCED (World Commission on Environment and Development) (1987) *Our Common Future,* Oxford University Press, Oxford.

Weale, A. (1992) *The New Politics of Pollution,* Manchester University Press, Manchester.

Winsemius, P. (1986) *Gast in eigen huis. Beschouwingen over milieumanagement,* Samsom H.D. Tjeenk Willink, Alphen aan den Rijn.

WRR (Netherlands Scientific Council for Government Policy) (1998) *Ruimtelijke ontwikkelingspolitiek,* nr. 53, Sdu Uitgeverij, Den Haag.

PART 2

ENVIRONMENTAL POLITICS IN AN ECONOMICAL CONTEXT

Chapter 4

GREENING PRODUCTION AS CO-RESPONSIBILITY

Walter J.V. Vermeulen

4.1 Introduction

Reducing the environmental impact of production processes in industry is traditionally one of the main targets of environmental policy. The government's role in this field is realised within a field of tension of differing expectations in society. On the one hand the freedom of the market economy must be maintained where possible; the creative potential of entrepreneurs must be used to the full; businesses must be treated equally; and existing 'rights' must be respected. On the other hand, radical changes in production processes are constantly called for, changes which – in the existing financial and economic structures – can be at the expense of earnings and the continuity of individual firms.

Policy pursued in the Netherlands in the 1990s to improve the environmental performance of businesses has since then undergone radical change. This has meant that the relationships between the government, the market and civil society have also changed considerably. The main actors in this process of change state quite emphatically that this policy is starting to bear fruit, to such an extent that we can now speak of a 'silent revolution' (Ministry of Housing, Spatial Planning and the Environment & VNO-NCW *(Confederation of Netherlands Industry and Employers)*, 1998).

This chapter sets out the trends we can observe in improving the environmental performance of businesses (the content) and in the societal process behind these improvements. We can see these trends as radical changes in the policy paradigm. With regard to the actual content, we detect

P.P.J. Driessen and P. Glasbergen (eds.), Greening Society, 67–90.

	industry's share in the total national emission		development in the total industrial emissions
	1990	1999	1990 - 1999
greenhouse gases	32%	33%	+ 9%
acidifying agents	19%	16%	- 43%
organic compounds:			
- CO	23%	22%	- 46%
- Benzene	9%	4%	- 77%
- Fluoranthene	55%	58%	- 39%
- Benzo(a)pyrene	54%	10%	- 93%
- Dioxin	4%	8%	- 88%
eutrophication (N-total)	20%	20%	+ 6%
eutrophication (P-total)	66%	40%	- 63%
waste produced	40%	36%	+ 3%
- recycled waste	45%	39%	+ 21%
- non-recycled waste	31%	27%	- 38%

Table 4.1 *Industry's share in the total national environmental load, the Netherlands 1990-1999 and the development in industry's total (% of the total for several emissions) (National Institute of Public Health and the Environment, 2000)*

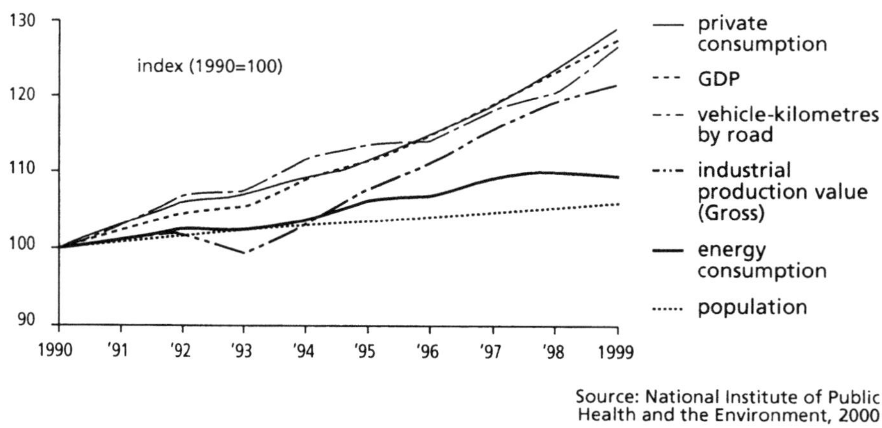

Figure 4.1 *Indicators of growth of production and consumption in the Netherlands*

a shift from static and singular goal-driven policies to dynamic and inclusive goal-driven policies. With regard to the societal process we see a shift from the central regulating state, as the guardian of the general (environmental) interests, to extended co-responsibility.

Before moving on to discuss the developments behind the claimed effectiveness of this renewed policy paradigm we first take a look at the actual environmental load caused by manufacturers. The shifts in content are then discussed and an outline is given of the way in which the extended co-responsibility is shaped by using a mix of policy strategies. Subsequently, we look at the extent to which the new concepts have penetrated the practical situation, i.e. whether they really have had an effect on the performance of businesses. The chapter concludes with a reflection on the driving forces behind the aforementioned transitions.

4.2 The producer's share in environmental pollution

That businesses in such a highly industrialised country as the Netherlands make the greatest contribution to the environmental problem goes without saying. Industrial production is responsible for one fifth to half of the emissions in most environmental themes. With regard to several emissions a great improvement was achieved in the 1990s (see Table 4.1), often a relatively stronger improvement than in other segments of society (agriculture, consumers, generation of electricity and transport).

Compared with that in other European countries, Dutch industry is relatively energy-intensive, yet the share of some emissions in the total national emissions is relatively lower (e.g. SO_2 and NO_x) (European Environmental Agency, 2000). Conversely, the environmental load in the Netherlands is relatively high because it is concentrated in a small area.

However, these data regarding industry's contribution towards the environmental problems only present a limited picture of the environmental load connected with manufactured goods and the importance of a further reduction of that load. Three comments are appropriate in this respect:

- Firstly, the share of the production stage in the environmental load caused by products is relatively low. The environmental load resulting from the extraction of raw materials, the utilisation of products and their disposal (post-consumer stage) are not included in the figures.
- Secondly, while we see a downward trend in certain emissions, even in a growing economy, we still see a distinct rise in others. The continued economic growth and the relatively high growth percentages in the field of transport and consumer spending (see Figure 4.1) can cause a U-turn in the decrease achieved so far and subsequently amplify the

increases. In the past, when formulating the goals for emission reductions a continued growth was seldom anticipated, while relatively high growth figures have still been achieved recently (and probably will be in the near future too).

- Thirdly, part of the environmental load from western prosperity, or in this case Dutch prosperity, occurs elsewhere. Globalisation contributes towards a part of the environmental load being shifted to other stages of the life cycle or to other countries or continents (see for instance Wackernagel & Rees, 1996; Carley & Spapens 1998; Van Vuuren et al., 1999).

4.3 The paradigm shift in the content of environmental policies

Major changes can be observed in the way in which the environmental load caused by industry is seen and, in line with this, in the way in which the long-term goals for improving the environmental performance of manufacturers are underpinned. As we already commented in the introduction, these shifts can be referred to as a development from static and singular goal-driven policies to dynamic and inclusive goal-driven policies.

The first shift is on the level of individual businesses and their relevant production chains. Up to the 1990s firms were addressed mainly on a series of separate emissions. Emission requirements were drawn up for each type of emission; emissions which basically could be traced back to ecological quality requirements and technological and economic considerations regarding emission-reduction technologies (usually end-of-pipe technologies).

Since the 1990s, the environmental performance of businesses has been increasingly assessed on the basis of its place in the life cycles of products. Scientists have come up with methods such as pollution prevention, life cycle assessment, life cycle management and integrated chain management for the systematic improvement of environmental performance. In other words, the relationship between environmental load and the effects of improvement options in the entire life cycle, for all relevant types of environmental load, can be analysed and assessed.

The development and application of these methodologies has been stimulated enormously by the Dutch Government. Intensive cooperation has evolved between the foremost knowledge institutes, the major umbrella organisations in trade and industry, and a select group of companies (often internationally operating companies). The search for new technologies also plays a major role. Not only have the technologies that can be applied immediately in the here and now, but also demonstration projects have been carried out. These should give rise to more faith in the possibilities for setting out a specific direction for the public and private efforts made in the field of technology development. This was the philosophy behind the *DTO,* the Sustainable Technology Development Project, a form of cooperation between the knowledge infrastructure, the government and industry's elite. Examples of far-reaching functional and system innovations have been outlined in the field of food, housing, mobility, chemicals and water, resulting in improvements in the eco-efficiency by a factor of 20 over a 50-year period (Bouwmeester & Jansen, 1997; Weaver et al., 2000). As a result of these – successful – demonstration projects, new initiatives have been started up focusing on radical technological innovations.

The second relevant shift relates to the level on which the collective environmental load generated by the individual firms is assessed. The essence of this shift is that long-term goals are no longer based on isolated national environmental problems, but are placed in a context of international justice.

Policy up to the 1990s was based on the aggregate of emissions in the different sectors in one's own country; used as the basis for reduction goals, usually with a time horizon of 10 to 20 years at most. Policy is also becoming more subjected to the influence of the so-called 'factor X discussion' (Reijnders, 1998). Publications in this field contain the message that an increasing level of prosperity *can* or *must* go hand in hand with a far-reaching reduction of the environmental load.

The currently dominant philosophy, in which the various approaches have been combined, can be described as a four-step route towards substantial improvement of product environmental performance (see Figure 4.2). Environmental management at the level of individual firms, resulting in an improvement in production processes, is the first step. The optimisation and redesign of existing products is the second. The greater part of environmental policy focuses on these two initial steps. The third step is to develop new forms of function fulfilment (e.g. e-mail instead of written letters). The fourth step consists of radical system innovations such as those illustrated in the sustainable technology development project.

On the one hand we are concerned here with arguments which – by making references to successes with pollution prevention, eco-design and system innovations – claim it is possible to achieve an eco-efficiency reduction by a factor of 4 or 10 in 25 years' time (Von Weiszäcker et al., 1997; Factor 10 Club, 1997). On the other hand, there are arguments which, on the basis of an analysis of the long-term growth expectations in terms of size of population and the level of prosperity in the whole world, come up with the normative stance that western economies must produce their consumption

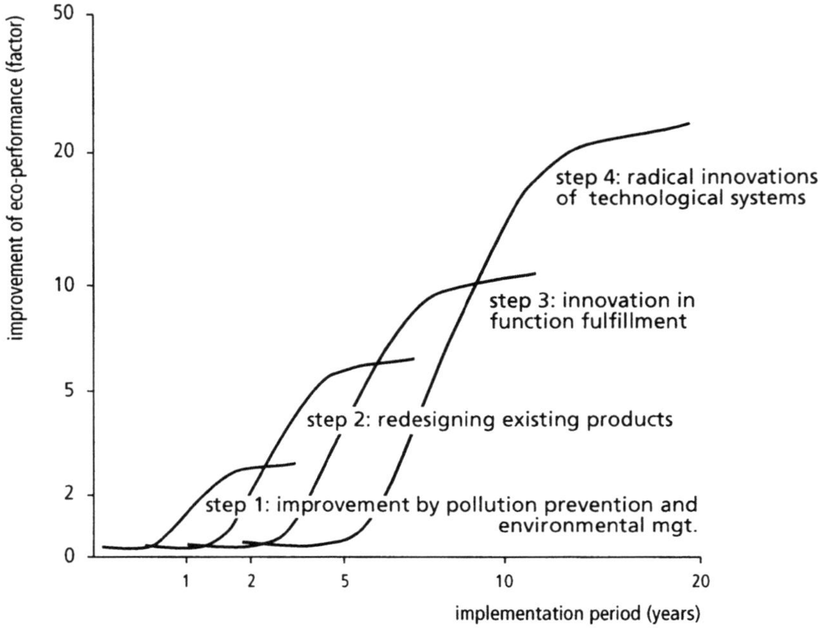

Figure 4.2 Four steps on the stairs of improvement of eco-performance

needs more efficiently in 50 years' time by a factor of 10, 25 or 40 (Weterings & Opschoor, 1994; Carley & Spapens, 1998; Raskin et al., 1998).

These discussions have had consequences for the environmental policy pursued in the Netherlands. The long-term goals have changed from:

- goals focusing on direct discharges;
- through goals formulated at the level of the total national emissions, linked to target groups;
- to ambitions formulated as an improvement of the (total) environmental performance per unit of product, taking into account the constant growth in level of prosperity.

The 'factor X discussion' also resounds in policy. The intention is to accomplish a fifty percent reduction in the total environmental load together with a doubling of the level of prosperity over the next 25 years. To achieve this so-called 'uncoupling' (between economic growth and environmental load) a great deal is expected in terms of new technological applications, the redesign of products and the use of renewable raw materials (such as sustainable sources of energy). The Dutch Government is searching for new ways to achieve 'long-term transitions' (Ministry of Housing, Spatial Planning and the Environment, 2001).

4.4 A paradigm shift in the process of environmental policies

The paradigm shift in content has emerged alongside a paradigm shift in the process of environmental policies. The two cannot be seen as separate. Developments in terms of content have helped to make developments in the process of environmental policies possible, and vice versa. In the 1990s the relationship between the state, the market and civil society in the Netherlands underwent radical change: from central management by the state to extended co-responsibility.

In the early years of environmental policy the solution for industrial pollution was searched for in a system of permits and licences. Faith was then placed particularly in the juridical model of management. Economic instruments were also used (including levies on fuel and the discharge of effluent). This strategy of central management via coercion and incentives failed miserably. In response, a mix of three management strategies was developed. There is still an element of central management, but in a contemporary form. Interactive management, combined with an internalisation strategy, was also emphasised (see Chapter 3). In this connection we have also seen forms of self-management emerge over the past few years.

Strategy 1: Central management by means of coercion and incentives
Modernisation by means of:
- reinforcing and relieving the implementation system;
- streamlining procedures;
- making more use of general rules;
- new forms of flexible and compact permits;
- greening the tax system.

Strategy 2: Interactive management and internalisation
- Agreements drawn up with industries in the framework of:
    - policy on target groups;
    - policy focusing on improving energy efficiency;
    - policy focusing on prevention and the reuse of waste;
    - policy on chemicals.
- Stimulating environmental management in companies.
- Extended producer responsibility.
- The development of methodologies.
- Information transfer and capacity building.

Strategy 3: Self-management
The stimulation of:

- privately formulated standards (ISO);
- recycling arrangements;
- entrance requirements for trade organisations;
- self-imposed environmental levies;
- responsible care;
- chain management;
- cooperation between manufacturers and consumers.

Strategy 1: Central management by means of coercion and incentives
Since the end of the 1980s work has been carried out on a more efficient juridical framework and on greening the tax system.

A more efficient juridical framework has been created by reinforcing the implementation system (more civil servants for the issue of permits and more inspectors) and the integration and streamlining of procedures in environmental laws. Moreover, an attempt has been made to relieve the implementation system by bringing a large number of businesses under the terms of general regulations, and thus businesses need now only 'report' instead of having to go through the permit procedure. In the 1990s, the administrative capacity for implementing environmental policy in both municipalities and provinces was more than doubled. Around 2000 it was said that there was sufficient implementation capacity, an adequate level of permits were being granted, and that an adequate number of inspections were being carried out.

The actual method of granting permits has also been changed. Whereas in many cases permits were drawn up as detailed voluminous documents, flexible permits are now granted under certain conditions. This is the case if a business is able to produce evidence that it has an appropriate (ISO 14000 approved) environmental management system in place. These permits set out the main lines only, the details being incorporated in the company's environmental plan and environmental programmes.

As we have already explained, in addition to the juridical instruments, economic instruments have always played a role in environmental policy. There is a significant trend underway in the direction of a green tax system. In 1993 a new law came into force in the Netherlands for a green system of taxation. For some time now tax has been levied on energy carriers and motor vehicles. Since then, green taxes have also been imposed on ground water and landfill waste, and work is under way on an expansion to include non-sustainable timber. The extent of environmental taxation is growing constantly (Figure 4.3) on the principle of 'impose tax on unwanted environmental pollution instead of essential labour and profits'. There is also a substantial increase in number and type of fiscal 'greening measures'. Examples are: making it an attractive proposition from a tax

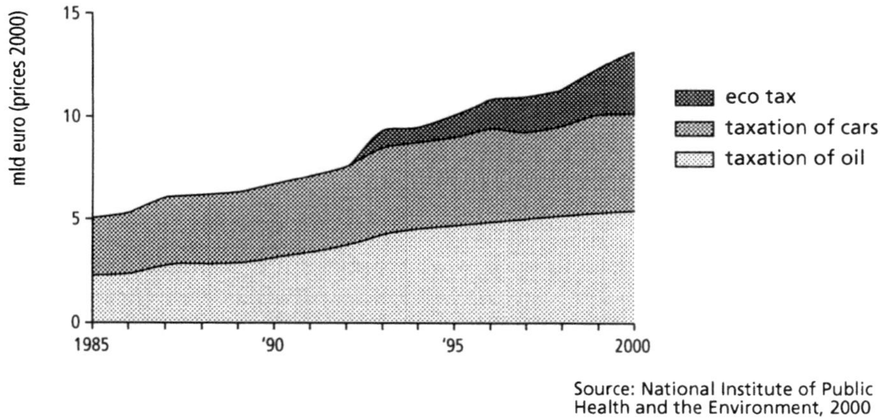

Figure 4.3 Greening national taxes in the Netherlands

point of view to make green investments, take out green mortgages and invest in green investment funds. The increase in the tax burden from environmental taxes is compensated by lowering taxes on income and profits for businesses and citizens pro rata. This now means that the Netherlands is among the countries with the highest share of 'green' taxes in the total package of taxes in Europe (European Environmental Agency, 2000).

In 2000, the national government's green taxes brought a total of EUR 13.2 million into the treasury. The share of green taxes in the total tax proceeds in the period 1992-2000 increased from approximately 9% to 15% (National Institute of Public Health and the Environment, 2000). Yet these figures still fail to provide a complete picture of the 'greening' because end-of-life fees were introduced on a variety of consumer goods in the 1990s to finance their reuse. These fees are not included in the national estimate.

In addition to these 'traditional' economic instruments, new instruments were recently introduced such as incentives for the leaders in the field, cost balancing among firms and benchmarking. With regard to these instruments, policy instruments are no longer linked to the individual company but to the mutual dynamism among the companies within an industry.

Strategy 2: Interactive management and internalisation
In addition to central management by means of coercion and incentives, a second management strategy was developed in the 1980s. This strategy consists of consulting with trade and industry and encouraging 'internalisation'. In this context, consultation focuses on reaching

agreement on the goals of national environmental policy and on working them out in greater detail together with the relevant industries. Internalisation aims at making these goals the 'goals of the company concerned' and at ensuring that the way in which the companies are organised makes it possible to achieve these goals.

Several adjacent paths were taken in the consultations with the industries:

- the *target groups consultation;*
- the consultation dealing with 'waste streams' on *pollution prevention and reuse policy;*
- the *long-term agreements* with the industries on the subject of energy efficiency.

The long-term targets incorporated in the first National Environmental Policy Plan were elaborated upon and specified for 16 target groups in the target groups consultation (such as chemical industry, metal industry, paper industy, printing industry, etc.). The agreements in this respect were usually made by the industry's trade organisations and the various government authorities. In several cases, companies were themselves responsible for signing the agreements. In these agreements the trade organisations were given a key role in specifying the technical requirements and monitoring the results.

The approach taken with regard to about 30 waste streams differed on several points. NGOs also played a role in these consultative talks. In applying the principle of producer responsibility so-called 'recycling arrangements' were settled. This included new legislation introduced at the request of trade and industry, among other things establishing a legal end-of-life fee – either in the product price or otherwise – to be paid by the manufacturer or the consumer. One step further than had been taken in the target groups consultation was taken in its implementation: organisations being set up for the collection and re-processing of waste.

The approach taken in the consultations with about 30 industries on improving *energy efficiency* is different again. It differs from the target groups consultation because in the preparatory stage several model companies were first visited in order to establish which technical possibilities were available for the improvement of energy efficiency. The results were then taken as the basis on which to draw up goals for the industries and to then lay them down in agreements which were signed by the companies individually. Subsidies and tax facilities for research, consultation and investment contributions were introduced systematically in the preparatory work and subsequent implementation.

Drawing up agreements with trade organisations will have little effect if, when conducting their business, the members themselves fail to focus on improving their own environmental performance. Running parallel with the consultations with the various industries, central government encouraged the introduction of 'in-firm environmental management systems' in companies in the 1990s. The roots of these management systems lie in the United States. All organisations ought to be structured in accordance with the ideal type of perfect management, the so-called Deming circle: *plan – do – check – react – plan.* This approach laid the foundation for an environmental management practice that also became the model for the introduction of in-firm environmental management systems here in the Netherlands. The organised elite of trade and industry played an encouraging role in this respect. In addition to the defensive character (preventing liability) it also assumed an offensive character in several large companies on the basis of practical experiences: systematically devoting attention to environmental aspects gives rise to financial savings: Pollution Prevention Pays (Huisingh et al., 1986; Dieleman & De Hoo, 1990; Schmidheiny, 1992).

The introduction of in-firm environmental management was encouraged by central government through an extensive programme of demonstration projects at branch of industry level, building up a network of consultants, developing modified forms of permit granting and initially by threatening to take coercive measures if the results were disappointing. Companies with a proper functioning environmental management system should subject themselves to checks in the form of external audits. The European Union introduced the EMAS regulation to this end and, since 1996, the international ISO 14001 standard has been used for the certification of environmental management systems. This standard stipulates specifically that all companies should work on the 'continuous improvement' of their environmental performance. Supervision has been privatised through a small number of accredited auditors and the cost of enforcement is therefore at the expense of the companies themselves.

Strategy 3: Self-management

The transition from the second to the third strategy is a gradual one. The third strategy implies stimulating self-activation among private organisations. Private parties ensure private regulation (within publicly established frameworks). Some remarkable examples of self-management have been seen over the past decade; many arising in response to government activities within the framework of the second strategy described above. One example is the trade association for companies engaged in the processing of chemical waste which, since the early 1990s, stipulated that a company must have an externally-audited environmental management system in place for it to be accepted into the association. This is of vital importance for companies given that clients may only do business with the members of this trade association.

The 'life-cycle settlements' resulting from the consultation on pollution prevention and reuse were also dealt with. The establishment of organisations such as 'Autorecycling Nederland' *(Car Recycling Holland),* for instance by those market parties involved in the reuse of wrecked cars, such as trade organisations for the car industry and car trade, is an example of how trade and industry generates new organisations for the purpose of achieving agreed upon goals. Collaboration with the government continues where necessary in this respect, for instance through the legal embedment of the end-of-life fee included in the purchase price of a new car (as is also the case in the purchase price of other consumer goods). Moreover, within the framework of the agreements on the reuse of car wreckages, reports must be submitted to the government on the progress achieved in realising the goals.

One radical form of self-management is seen in the ISO standards mentioned in the previous section; standards which continue to play an increasingly important role in environmental management. Originally, the definition of the requirements for environmental management systems were drawn up by the government (Ministry of Housing, Spatial Planning and the Environment, 1989). Because of the similarities with the quality management systems the government adopted the ISO standard for quality management systems as the standard in the 1990s. This was in response to demands made by the market. Meanwhile, the International Standardisation Organisation (ISO) had started work on the standard for environmental management systems, leading to the international series of standards, ISO 14000. Standards originating from ISO are a perfect form of self regulation. The role government plays in establishing the standard is, at most, the role of one of many interested parties. There is no control by parliament or any parliamentary contribution towards realising standards. Moreover, once a standard has been established, its application is through private channels. Verification as to whether products and organisations meet the requirements is carried out by authorised commercial management consultancies. However, the government does play a role at arm's length in different ways. On the one hand it creates the organisational framework by means of the legal embedment of the 'Nederlandse Normalisatie Instituut' (NNI) *(Netherlands Standardization Institute)* and the 'Raad voor Accreditatie' *(Dutch Council for Accreditation).* On the other hand, formal regulations and standards, and the goals formulated in government policy documents, are used as reference points in the analyses companies must make and must take into account in their policies in accordance with the various requirements specified in ISO 9000 and ISO 14000 standards (ISO, 2000; SCCM, 2000).

It is quite feasible that it is thanks to the private nature in particular that the ISO standards for quality management and environmental management have started to have a converging effect in communication on the environment in the market (particularly in the relationships between companies, and possibly in the future more in the communication between manufacturer and consumer). This private regulation is starting to play a major role in the relationship between companies and the government too, especially in connection with the question whether a company is eligible for a flexible permit.

4.5 The shifting role of technology policy

The orientation towards the long-term task of achieving a substantial improvement in efficiency as to how we use our energy and materials (described in section 4.3) also brings technology policy into the picture. This was not dealt with in the strategies discussed above because of the fact that for some time now environmental technology policy has been evolving quite separately from environmental policy. The development of environmental technology at universities and the major national research institutes has received financial support, even from the early years of environmental policy. There has also been a large number of subsidy regulations in place over the years to support technology development and the introduction of new environmental technology.

Technology policy also focuses on the acceleration and success of market introduction. After the first breakthrough of a new technology on the market, further diffusion must be coaxed by way of regular policy focusing on the business community. It is then a case of quickly bringing an innovation up to the level of 'state-of-art technology'. This can be achieved by including an innovation on the list of technologies eligible for tax advantages, by including it in the manuals issued by the licensing authorities, and by using these innovations as the basis for agreements with industries. This 'process of effecting society' usually involves a certain amount of delay.

The growing level of understanding into how the innovation diffusion processes progress makes it quite clear that a policy geared towards technology push, by giving financial support to research and development alone, is not enough. Successful innovations involve close contact between innovators and users. In this context reference is made to reinforcing clusters and collaborating businesses (Jacobs, 1990, Roelandt et al., 1997; OECD, 1992). The barriers standing in the way of market introduction must also be eliminated (Cramer & Schot, 1990). All of this calls for the sound coordination of government's efforts in the various stages of the process of development, market introduction and wider diffusion.

To an increasing extent are attempts also made to activate 'innovation networks' on the basis of a long-term perspective by means of 'strategic conferences', 'taskforces' and 'chain consultation'. Technology foresight studies, used to identify promising technologies on the basis of the long-term targets for sustainable development, also play a similar role (see National Institute of Public Health and the Environment, 1994; TNO & CPB, 1997. Stimulating innovation networks is done by granting subsidies to joint ventures, the producers of environmental technology and the intended users only. Examples of these subsidy programs in the Netherlands are the 'Economy, Ecology and Technology' program and the 'National Initiative for Sustainable Development' program. These programs focus on projects that are able to achieve a technological breakthrough in 5 to 20 years time or even longer, being orientated towards the 'factor X discussion' referred to in the foregoing.

4.6 Shifting paradigm, bending curves?

The mix of management strategies discussed above was developed in the 1990s. In the introduction we already mentioned the consensus among the main actors regarding the success of the route taken. To what extent does this approach already lead to a dislocation in the trend of how Dutch society meets its material needs. Is there really a shift away from the trend? This question can be answered on the basis of physical science indicators and social science indicators. We will be brief in looking at the former and devote most of our time to look at the information concerning the extent to which changes arise – partly as a result of policy – in the social organisation of production, and whether this leads to the desired shift in practice.

Partly reduced emissions
Studies carried out by the National Institute of Public Health and the Environment (RIVM) show that over the past decade there has been a partial uncoupling (Figure 4.4). The current ongoing economic growth is accompanied with improvements in waste disposal and recycling, acidification and soil cleaning. There is hardly, if any, uncoupling in terms of climate change and disturbance (National Institute of Public Health and the Environment, 2000).

To obtain more insight into the societal changes concealed behind these scientific data we must take a look at the social science research carried out on this subject. To what extent have the three strategies – central management, interactive management and self-management – led to an improvement in environmental performance?

Whereas there is a large amount of research available in this field in the Netherlands, the research agenda itself tends to be open to fashions. Of note is that there has been very little research carried out into effects of the central management strategy by means of coercion and incentives on industry. Conversely, the strategy of interactive management and

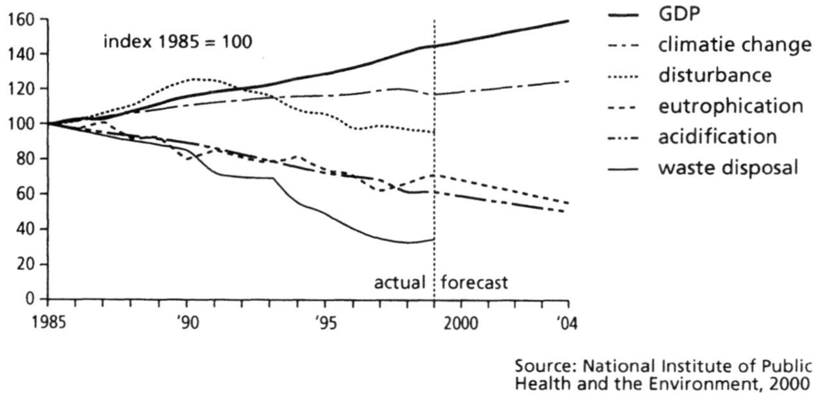

Figure 4.4 *Performance indicators of some environmental policy themes and GDP*

internalisation has been given more attention. We shall discuss research into environmental management, voluntary agreements, self-management and product design.

Environmental management

The introduction of environmental management is recognised by many as a success. With a view to the original policy objective ('all the companies that caused the most pollution have an in-house environmental management system in place in 1996') it can be said that the success was partial, yet there is still an ongoing diffusion of environmental management systems. The number of businesses in which environmental management has become routine, or in which implementation of such a system is under way, has grown from 37% in 1997 to 49% in 1999 (see Figure 4.5; KMPG Milieu & NIPO, 1999: 24; KPMG Milieu & IVA, 1996). However, there are differences to be seen among the various branches of industry. The leading sectors are building materials, chemical, paper and foodstuffs (60-70% in the routine or implementation stage). Lagging behind is the building trade, road transport and electrical engineering (less than 45%).

While these figures are indicative of a high diffusion level of environmental management systems, the underlying data were obtained from surveys and give too little insight into the actual effects of environmental management systems on management. Conversely, other studies show a limited impact of environmental management systems on the day-to-day running of businesses. An important role is attributed to external influences on business strategy. If these are lacking, then

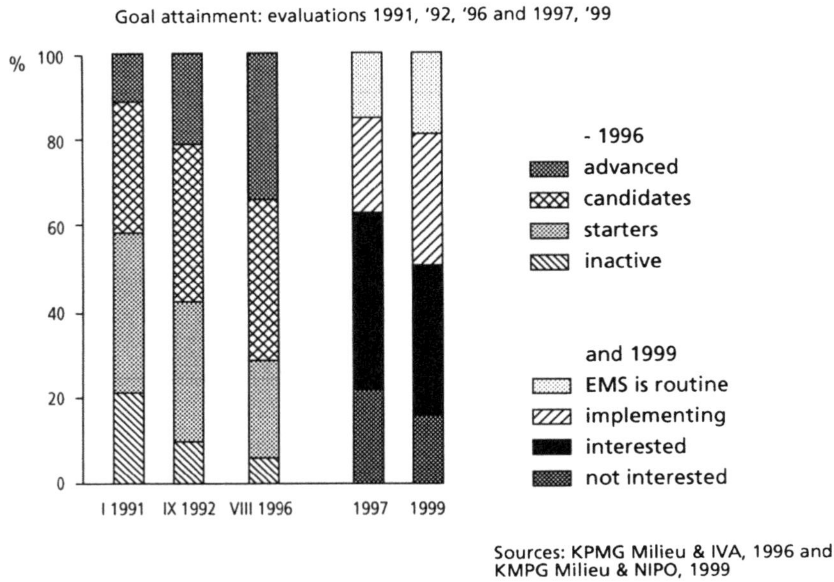

Figure 4.5 Implementing environmental management systems

businesses can easily engage in environmental management *pro forma* without this in-firm environmental policy really penetrating through to the company's policy (Bouma, 1995, p. 481, 520; Le Blansch, 1996, p. 287). Moreover, it is also observed that the external stimuli in the first half of the 1990s were very weak (De Groene, 1995). For instance it was noted that the environmental consultancy structure for businesses was highly fragmented and consequently lacked effect (EIM, 1994). Moreover, in the mid-1990s only 46% of the municipal authorities actively encouraged businesses to introduce environmental management systems. It is quite remarkable to see that some of these municipal authorities (38%) were not only passive in this respect, but also put up resistance against this national policy (De Bruijn & Lulofs, 1996, pp. 408-409). In response, central government is currently conducting demonstration projects to show businesses how to incorporate environmental policy into their own strategic company policy.

Voluntary agreements

Agreements with target groups on the subject of environmental policy, recycling and energy, play a pivotal role in the (second) strategy of interactive management. Policy focusing on the setting up of a return logistics for waste streams is visibly paying off. For instance: figures published by the National Institute of Public Health and the Environment show that a radical shift was achieved in the methods used for processing during the period 1990-1999. The percentage of waste that is either reused or given a useful application has increased from approximately 50% in 1985 to 76% in 1999. The objective was 80% in 2000. The amount of waste dumped in that period was brought down from 28% to 10% (National Institute of Public Health and the Environment, 2000).

In 1998, central government and the employers co-published the book entitled 'De stille revolutie' *(The Silent Revolution)* in praise of their 'joint action to achieve a better environment'. In this book they state that of the 73 industrial emissions, the intended reductions for the year 2000 were achieved in 51 cases as early as 1996, that the targets for 2000 would be reached for 11 emissions, and that current policy was inadequately effective for 11 emissions only (Ministry of Housing, Spatial Planning and the Environment & VNO-NCW, 1998, pp. 160-161).

Glasbergen places successes such as these against the background of the discordant relationship between government and trade and industry in the previous periods and characterises the target groups consultation, and the role of the agreements therein, as a learning process for all concerned: government, trade and industry and the environmentalist movement. Agreements in this respect function as a focal point on the communal route that leads towards cooperative environmental management (Glasbergen, 1997; 1998a and b; 1999). This positive view is also seen in the evaluation study of the Long-term Agreements on energy efficiency. Despite the low price of energy, businesses are more highly motivated, have obtained a better insight, and much more serious thought is given to improving energy efficiency (Glasbergen, 1997).

Voluntary agreements with trade and industry are experienced more positively in the Netherlands than in many other countries because the approach taken in the Netherlands is less non-committal, monitoring and feedback is generally well-organised and firm links have been established with the other instruments (see also Environmental Law Network International, 1998; OECD, 1999).

Self-management

In section 4.4 we discussed the ISO standards as an example of self-management. The extent of this standard's application is probably the most absolute indicator of the level of environmental management internalisation, much better than the surveys on environmental management mentioned above. The recurrent intensive examinations carried out by the auditor are aimed specifically at finding out whether the organisation functions in accordance with the principles of the 'plan-do-check-act cycle' and 'continuous improvement'.

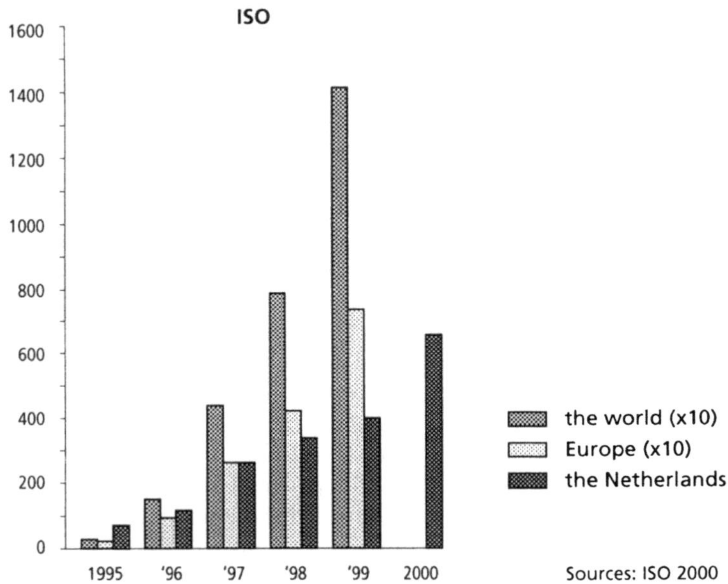

Figure 4.6 *Developments in certification of environmental managmentsystems*

While the ISO 14001 standard was introduced in the Netherlands in 1996, the expectation is that a similar growth will be seen for this as we saw for the diffusion of the certification of quality management (ISO 9000) (EIM, 1997, p.87). However, at the beginning of 2000 only about 650 businesses in the Netherlands were in possession of an ISO 14001 certificate (ISO, 2000). This means that the Netherlands is certainly not one of the leaders in this field. Almost fourteen thousand ISO 14001 certificates were granted world-wide at the beginning of 2000 (i.e. within a period of four years), half of which were in Europe. Businesses in the far east are also active in this respect. The sectors in which certification is the most successful are electrical and optical equipment, chemical, and the manufacture of transport equipment (ISO, 2000).

Product redesign
We saw in section 4.2 that to achieve radical environmental improvements a shift must be achieved from the improvement of individual parts of production processes to an integral environmental assessment of production chains and products (by using methods developed for integral chain management and environment-oriented product development, etc.).

To this end, it is essential that manufacturers join forces in, *inter alia,* the field of technology development and the further diffusion of environmental innovations. Research into 'cleaner manufacturing' provides insight into the sort of measures taken by manufacturers. Depending on the sort of measure concerned, in 1999 between 25 and 65% of all businesses state to have introduced specific measures over the previous two years. The emphasis is apparently on the use of base materials with a lower environmental impact, energy management and modified working methods. Prevention checks (26%) and environment-oriented product development (25%) have made less headway to date (KPMG Milieu & NIPO, 1999).

A similar picture emerges from a European study into the use of eco-design methods in trade and industry. One of the conclusions of this study is that even in those countries that led the way in the development of methodologies, including the Netherlands (Brezet, 1994; Brezet & Van Hemel, 1997), their use in practice remains limited to a select group of large, internationally operating companies (Tukker et al., 2000). It is quite possible that a relatively small group of businesses is emerging which is able to demonstrate the feasibility of far-reaching improvements in environmental performance. Nevertheless, this group fails to connect up with the other businesses where 'business as usual' prevails.

This conclusion is quite remarkable to say the least. There have been many win-win situations since the mid-eighties regarding the prevention of waste and emissions. Moreover, this message has been brought to the attention of trade and industry quite intensively since 1990 through pollution prevention demonstration projects and the establishment of provincial 'pollution prevention teams' (see e.g. Dieleman & De Hoo, 1991; Van Berkel, 1996). If the effect of this pleasant message is so difficult for this theme, what then is the situation regarding recent, more complex aspirations such as integral chain management (saying: "co-operate with your suppliers and clients") and environment-oriented product development (saying: "assess your product properly and conquer new markets")? Research conducted into the experiences in trial projects with product-oriented environmental management systems (as an integration of product policies in environmental management systems) also shows a distinction between a leading group of large, or conversely, small businesses (which have made improving environmental performance the core element of their business policy) and other businesses (Brezet et al., 2000). It had been observed before then that the successful effect of ecodesign advice in leading businesses, for instance, could be explained from the presence of an environmental management system and the willingness to innovate in the businesses concerned (Van Hemel, 1998, p. 232). Other research, however, shows potential breakthroughs among the 'followers': in 1999, 68% of all businesses planning to market new products say they will devote attention to the environmental aspects in the development process (KPMG Milieu & NIPO, 1999, p. 35).

Both the tendency towards interactive management and self-management, as well as the efforts to collaborate on technology development, calls for more intensive relationships. Businesses will need to work together with other parties in the market more often, parties such as suppliers, customers, trade organisations and knowledge institutes. This increasing level of collaboration among the market parties is not only seen in the Netherlands. International 'environment-conscious' employers' organisations, such as the World Business Council for Sustainable Development (WBCSD) and the Coalition for Environmentally Responsible Economies (CERES), play a significant role in the international widening of these developments. Much of the practical examples used in the discourses are drawn from these companies. This supports us in drawing the conclusion that an internationally-oriented business elite is to a large extent steering the development towards extended co-responsibility.

4.7 Reflection: co-responsibility sufficiently penetrated?

Considering the studies discussed above into the effects of the different management strategies, the reduction in environmental load that has been realised must be attributed mainly to the second strategy: interactive management and internalisation. This has resulted in records showing definite successes with a wide bandwidth. Regarding the paradigm shift in terms of content, for the time being we are only able to record any effects among a small group of leaders. It is expected that a further, substantial uncoupling of the growth in prosperity and the environmental load will be achieved in the future. The question is, however, what driving forces will be behind these processes of change.

Some theorists state that there is talk of a process of 'ecological modernisation', a process which in terms of impact would be comparable with the Enlightenment in the 16th century, the process of State Formation in the 17th and 18th centuries, the Industrialisation in the 18th and 19th centuries, and the Social Emancipation in the 20th century (Huber, 1991). While some take a more analytic-sociological perspective, often discourses on ecological modernisation tends to take a normative perspective. The subject under discussion is the changing relationship between the state, the market, science and technology and civil society (Huber, 1991; Mol, 1995; Jänicke, 1993). Some publications focus on the changing role of the state: from curative to preventive, from closed policy making to participatory, from dirigistic to contextually steering (Mol, 1997, p. 141). Others lay a greater emphasis on the new dominant role of trade and industry. For instance, Mol (1995, p. 36, 58) (referring to Huber, 1991) states that modern science and technology, the economic actors and market mechanisms play a leading role in the process of ecological transformation, while the government, political and societal movements are becoming less of a driving force behind this transformation. The link Huber makes with the 'economic long-wave theory' gives ecological modernisation an almost unavoidable character, based on a nearly blind faith in the market forces and technology (see for instance Huber, 1991, p. 177).

Mol states quite rightly that, when formulated in this way, the epistemological problem arises of how to evaluate the value of a social theory that is so closely related to and derived from social practices and dominant ideas (Mol, 1995, p. 57). In the foregoing we have explained how the changes in terms of content and process were realised in close collaboration between the government, institutes of knowledge and a limited number of companies, labelled here as a business elite. While the role of organisations in civil society in these developments has been an alternating one, it was relatively modest.

The new cooperative relationships between state, market and knowledge institutes should not, however, be seen as a far too obvious guarantee for a successful 'ecologising of the economy' (cf. Huber, 1991 and Jänicke, 1993). The diffusion of new concepts and practices, such as life-cycle management, ecodesign and the like, are apparently for the time being not progressing as matter of course as the claimed ecological and economic gain would have us believe. A small group of leading companies do use them. Some dissemination can be observed among representing organisations and newly established players (like environmental consultancies, information transfer organisations and organisations in the field of life cycle cooperation and monitoring). Theses organisations even take their own stimulating role in this field. But these developments have simply passed by many 'run-of-the-mill' businesses.

It is quite possible that we are witnessing a partial greening of society, in which a well-organised internationally-oriented business elite is successful in conceptualising the intended transitions and thus making a strong impression on policy programmes. Simultaneously, for a large group of smaller, more nationally-operating businesses, this green entrepreneurship is too far away. However, societal processes of change will always be driven by a leading group of innovative actors. The risk involved here is that an ambitious long-term orientation gives too little consideration to the problems of the further diffusion of 'green entrepreneurship' to the wide segment of average businesses.

References

Brezet, H. (1994) *Van prototype tot standaard. De diffusie van energiebesparende technologie*, Rotterdam.

Brezet, H. & C. van Hemel (1997) *Ecodesign. A Promising Approach to Sustainable Production and Consumption,* UNEP, Paris.

Brezet, H., B. Houtzager, R. Overbeeke, C. Rocha & S. Silvester (2000) *Evaluatie van 55 PMZ-subsidieprojecten,* Den Haag.

Bouma J.J. (1995) *Milieuzorg bij de koninklijke landmacht en industrie. Een onderzoek naar de integratie van milieuaspecten in strategische besluitvormingsprocessen,* Rotterdam.

Bouwmeester, H. & L. Jansen (1997) *2040-1998: technologie, sleutel tot een duurzame ontwikkeling,* Interdepartementaal Onderzoekprogramma Duurzame Technologische Ontwikkeling, Den Haag.

Carley, M. & Ph. Spapens (1998) *Sharing the World: Sustainable Living and Global Equity in the 21st Century,* Earthscan Publications Limited, London.

Cramer, J. & J. Schot (1990) *Problemen rond innovatie en diffusie van milieutechnologie,* Raad voor Milieu- en Natuuronderzoek, Publicatie nr. 44, Rijswijk.

De Bruijn, T.J.N.M. & K.R.D. Lulofs (1996) *Bevordering van milieumanagement in organisaties,* Twente University Press, Enschede.

De Groene, A. (1995) *Beheersen of beïnvloeden. De respons van bedrijven op milieuproblemen. Het belang van de omgeving,* Middelburg.

Dieleman, H. & S. de Hoo (1991) *Kiezen voor preventie is winnen: naar een preventief milieubeleid van bedrijf en overheid,* NOTA, Den Haag.

EIM (1994) *Effectief stimuleren van milieuzorg bij kleine bedrijven,* Zoetermeer.

EIM (1997) *De innovativiteit van de Nederlandse industrie,* Zoetermeer.

Environmental Law Network International (1998) *Environmental Agreements. The Role and Effect of Environmental Agreements in Environmental Policies,* Darmstadt.

European Environmental Agency (2000) *Environmental Signals 2000,* Copenhagen.

Factor 10 Club (1997) *The International Factor 10 Club's Statement to Government and Business Leaders: A Tenfold Leap in Energy and Resource Efficiency,* Wuppertal Institute, Wuppertal.

Glasbergen, P. et al. (1997) *Afspraken werken, evaluatie meerjarenafspraken over energie-efficiency,* Universiteit Utrecht, Utrecht.

Glasbergen, P. (1998a) 'Partnership as a Learning Process', in: P. Glasbergen (ed.) *Co-operative Environmental Management: Public-Private Agreements as a Policy Strategy,* Kluwer Academic Publishers, Dordrecht.

Glasbergen, P. (1998b) 'Learning to Manage Energy by Voluntary Agreement: the Dutch Long-Term Agreements on Energy Efficiency Improvement', *Greener Management International,* 22, pp. 46-61.

Glasbergen, P. (1999) 'Tailor-Made Environmental Governance: On the Relevance of the Covenanting Process', *European Environment. The Journal of European Environmental Policy,* vol. 9, no. 2, pp. 49-58.

Huber, J. (1991) 'Ecologische modernisering: weg van schaarste, soberheid en bureaucratie?', in: A.P.J. Mol, G. Spaargaren & A. Klapwijk (red.) *Technologie en milieubeheer. Tussen sanering en ecologische modernisering,* Sdu Uitgeverij, Den Haag.

Huisingh, D, et al. (1986) *Proven Profits From Pollution Prevention. Case Studies in Resource Conservation and Waste Reduction,* Institute of Local Self Reliance, Washington.

ISO (2000) *The ISO Survey of ISO 9000 and ISO 14000 certificates: The Ninth Cycle,* Geneve.

Jacobs, D. (1990) *The Policy Relevance of Diffusion,* Ministry of Economic Affairs, Den Haag.

Jänicke, M. (1993) 'Über Ökologische und Politische Modernisierungen', *Zeitschrift für Umweltpolitik und Umweltrecht,* no. 2, pp. 159-175.

KPMG Milieu & IVA (1996) *Evaluatie Bedrijfsmilieuzorgsystemen 1996,* Den Haag/Tilburg.

KMPG Milieu & NIPO (1999) *Schoner produceren in Nederland 1999,* Ministerie van VROM, Den Haag.

Le Blansch, K. (1996) *Milieuzorg in bedrijven; overheidssturing in het perspectief van de verinnerlijkingslijn,* Utrecht.

Ministry of Housing, Spatial Planning and the Environment & VNO-NCW (1998) *De stille revolutie: industrie en overheid werken samen aan een beter milieu,* Distributiecentrum VROM, Den Haag.

Ministry of Housing, Spatial Planning and the Environment (Ministerie van VROM) (1989) *Nationaal Milieubeleidsplan. Kiezen of verliezen,* TK 21137(1-2), Den Haag.

Ministry of Housing, Spatial Planning and the Environment (Ministerie van VROM) (2001) *Nationaal Milieubeleidsplan 4. Een wereld en een wil; werken aan duurzaamheid,* Den Haag.

Mol, A.P.J. (1995) *The Refinement of Production: Ecological Modernization Theory and the Chemical Industry,* Jan van Arkel, Utrecht.

Mol, A.P.J. (1997) 'Ecological Modernization: Industrial Transformations and Environmental Reform', in: M. Redclift & G. Woodgate (eds.), *The International Handbook of Environmental Sociology,* Edward Elgar Publishing, Cheltenham, pp. 138-149.

OECD (1992) *Technology and the Environment. The Key Relationships,* Paris.

OECD (1999) *Voluntary Approaches for Environmental Policy. An Assessment,* Paris.

Raskin, P. et al. (1998) *Bending the Curve. Towards Global Sustainability,* SEI, Stockholm

Reijnders, L. (1998) 'The Factor X Debate: Setting Targets for Eco-efficiency', *Journal of Industrial Ecology,* no. 1, pp. 13-22.

National Institute of Public Health and the Environment (RIVM) (1994) *Technologische Oplossings Richtingen Keuzedocument (TORK),* Bilthoven.

National Institute of Public Health and the Environment (RIVM) (2000) *Milieubalans 2000,* Bilthoven.

Roelandt, Th.J.A., P. den Hartog & D.F.M.F. Jacobs (1997) *Nederlandse clusters in beeld,* Ministerie van Economische Zaken, Den Haag.

TNO & CPB (1997) *81 mogelijkheden; technologie voor duurzame ontwikkeling,* Delft.
Schmidheiny, S. (1992) *Changing Course: A Global Business Perspective on Development and the Environment,* MIT Press, Cambridge MA.
SCCM (2000) *Certificatiesysteem milieuzorgsystemen volgens ISO 14001,* R000616, Den Haag.
Tukker, A., E. Haag, & P. Eder (2000) *Eco-design: European State of the Art Part I: Comparative Analysis and Conclusions,* Institute for Prospective Technological Studies, Sevilla.
Van Berkel, C.W.M. (1996) *Cleaner Production in Practice,* IVAM, Amsterdam.
Van Hemel, C.G. (1998) *Ecodesign Empirically Explored. Design for Environment in Dutch Small and Medium Sized Enterprises,* Delft.
Van Vuuren, D.P., E.M.W. Smeets & H.A.M. de Kruijf (1999) *The Ecological Footprint of Benin, Bhutan, Costa Rica and the Netherlands,* RIVM, Bilthoven.
Von Weiszäcker, E.U., A.B. Lovins & L.H. Lovins (1997) *The Factor Four,* Earthscan, London.
Wackernagel, M. & W. Rees (1996) *Our Ecological Footprint. Reducing Human Impact on the Earth,* New Society Publishers, Gabriola Island.
Weterings, R. & J.B. Opschoor (1994) *Towards Environmental Performance Indicators Based on the Notion of Environmental Space,* RMNO, Rijswijk.
Weaver, P. et al. (2000) *Sustainable Technology Development,* Greenleaf Publishing Ltd.

Chapter 5

THE STUBBORN CONSUMER

Walter J.V. Vermeulen

5.1 Introduction

Environmental effects resulting from production and consumption are closely linked. An increasing level of prosperity, shifts in consumer patterns, and the alleged power of the consumer through the market brings the role of civil society into the picture in the transition to a sustainable society. The role of consumers and their organisations can – by way of an articulated demand for products with a lower environmental load – be a significant motive for manufacturers to accept the extended co-responsibility described in the previous chapter. Looked at from this point of view it is relevant to find out which trends are taking place in consumer behaviour, and what are the environmental consequences of those trends. It is also interesting to analyse how the interaction between market and civil society in this field takes shape, and what role is and can played by the state to influence consumer behaviour.

The growth in prosperity in the West brings a more normative question into the picture. In the Declaration of Rio, Agenda 21 dating from 1992, the United Nations called upon all states to impose restrictions on and to eliminate non-sustainable methods of production and consumer patterns. This, it was stated, was due to the fact that while there is an exceptionally high level of consumption and non-sustainable consumer patterns in the wealthier parts of the world, there is still a large section of mankind that is withheld from the most elementary consumer needs (UN Conference on the Environment and Development, 1992, pp. 8, 42-49). Reasoning from this perspective, the UN brought the question of the environment and

P.P.J. Driessen and P. Glasbergen (eds.), Greening Society, 91–111.

consumption into the context of international justice and solidarity. In doing so, the world's highest administrative body took an enormous step forward, a step not often taken at the level of the individual states. The question is in how far can this normative approach be brought into practice and what is the role civil society can play therein.

This chapter deals with the conscious attempts to reduce the environmental load caused by consumption. First, we take a look at the environmental effects of consumption, setting out a brief analysis of the various consumer patterns and look at the aspect of 'over-consumption'. We then take a look at the role of the state and discuss the effects of the policy pursued in this respect. Subsequently, we deal with the question of what can be expected from calls to change consumer patterns and values and discuss developments in the division of roles between the market, the state and civil society. We conclude by looking at the possibilities of influencing consumer behaviour.

5.2 The consumer's share in environmental degradation

Consumption and consumer patterns have only recently been given serious thought in environmental policy. We can regard this recent attention mainly as a reaction to the observed stubbornness of existing consumer patterns. Changes easy to achieve in environmentally relevant behaviour (such as separating waste, saving energy and water, which does not cost money, the latter two can even save money) have to a large extent been realised in the Netherlands. A substantial change in behaviour in these fields has been observed. The separation and re-processing waste is a success. Water consumption per capita is stabilising after having risen for some years in line with the growth in prosperity. What remains are environmentally relevant consumer patterns that are at the expense of comfort, status and luxury, or seem inevitable and consequently difficult to change. To what extent do these attitudes contribute towards the environmental problems? Table 5.1 shows that the environmental load in terms of themes such as climate change, acidification and waste is only a fraction of the total environmental load in the Netherlands. In other fields, the share of consumption in the total environmental load is much greater, sometimes because a 'major clean-up' has been held in industry.

The above figures relate to the environmental load generated in the use stage and the load that arises when the associated waste is removed. However, we must point out here that these figures only give a limited picture of the environmental impact that results from consumption. We obtain a much better picture if the environmental load is taken into

	consumer's share in the total national emission 1990	1999	development in the total consumer's emissions 1990 – 1999
greenhouse gases	10%	9%	- 2%
acidifying agents	3%	4%	- 1%
organic compounds:			
- benzene	6%	8%	- 26%
- fluoranthene	18%	19%	- 32%
- benzo(a)pyrene	28%	61%	- 21%
- dioxin (excl. incinerators)	5%	71%	- 19%
eutrophication (N-total)	65%	68%	+ 9%
eutrophication (P-total)	30%	55%	+ 9%
waste produced	12%	15%	+ 34%
- recycled waste	3%	9%	+ 278%
- non-recycled waste	27%	34%	- 12%

Table 5.1 *The consumers' share in the total national environmental load in the Netherlands 1990-1999 and the development in the total consumers' share (% of total for several emissions) (National Institute of Public Health and the Environment, 2000).*

consideration throughout the entire life cycle of the goods consumed. Individual life cycle analyses and studies into the *indirect* energy consumption, or how goods make use of the environment, show quite clearly that a far greater claim is made on the environment. Studies conducted by Vringer et al., (1997) show, for instance, that the indirect energy demand (i.e. the energy required to produce and transport consumer goods) is more than half of the total energy demand. More than two thirds of the total direct and indirect energy consumption is swallowed up by activities that relate to housing (39%) and caring activities (nutrition, care and clothing: 30%). The activities that 'make life pleasant' (recreational activities such as sport, amusement, development, etc.) make up for 18% of energy consumption. And last of all, transport takes up a considerable share, standing at 13%.

These figures only relate to energy and relate to random figures at a given moment in time. Over the last few decades we have seen a steady growth in household consumption in the Netherlands (measured in the total amount of Euros spent): over the past two decades this averaged an annual 1.75% and the growth was relatively high in the second half of the last decade of the 20th Century: 6%.

Not only consumption per capita or per household increased, the size of the population is also on the increase by about 0.7% per annum. Around the turn of the century we welcomed the six billionth inhabitant of Earth, and the 16 millionth inhabitant of the Netherlands. In the course of the 21st Century the population of the Netherlands is expected to stabilise at 17 million. This Dutch population is, however, starting to live in smaller households. The average thinning out of households over the previous decade was 1.6% per annum.

Such figures on more inhabitants, smaller households, increasing consumer expenditure, say very little in themselves on the actual size of the environmental load. The extent of improvement in environmental performance of consumed goods is also important. An attempt was recently made to estimate the effect of developments in the field of demography, behaviour, prosperity and cleaner technology. These analyses show that technological developments in the period 1985-1997 have led to a 15% reduction in the total indirect and direct energy demand of households. Conversely, there is a 9% increase due to population growth, an increase of 4% because of the smaller households, an increase of 9% due to the rise in income and, last but not least, an increase as high as 18% as a result of behavioural trends (National Institute of Public Health and the Environment, 1998).

Technological environmental improvements are occasionally counteracted by gradual changes in behaviour. One example in this respect is the change in how we do our laundry. Improvements in detergents and in washing machines have made it possible to wash at lower temperatures and thus achieve a considerable saving in energy. This saving has almost been negated by to the fact that Dutch people have gradually started to wash more items with a higher frequency over the past few decades. The total energy consumption for laundry purposes has remained stable (Slob et al., 1996).

A long-term foresight study into the use of energy for consumption purposes shows that, while an improvement in energy efficiency can be expected over the period 1995-2030 in the region of 60%, we must still take into account an increase in the total use of energy for consumption by 50 to 75% (Vringer et al., 2001). Such figures stress the importance of anticipating the growth in prosperity when establishing environmental policy goals.

5.3 Analysing consumer patterns

To establish the environmental load on the basis of consumption we need data on the extent to which consumption shifts in the direction of fewer or less environmentally burdensome activities. To do this we must first have an understanding of how money is spent on the various categories of expenditure and information about the environmental load of those categories. An increasingly smaller part of our incomes is being spent on nutrition and clothing (approximately 20% in Europe). These are general trends in the western world. While expenditure in the field of *housing* shows a decreasing tendency in the United States, in Europe it is stable (17-18%). In western nations there is a gradual growth in the share of expenditure *on amusement and recreation* up to approximately 11% in Japan and the United States, and 9% in Europe. While this growth progresses slower in Europe, it is relatively strong in the Netherlands (Eurostat, 1998).

Does an increase in the level of prosperity lead to a higher environmental load? One way of finding out is by checking whether higher incomes lay higher claims on the environment. One such analysis was carried out by Vringer et al., (1997) by analysing the direct and indirect energy consumption of the Dutch population divided into ten equally large groups with incremental incomes (the so-called incomes percentiles: the first percentile is the 10% of households on the lowest incomes, the tenth percentile is the 10% of households with the highest incomes). The analysis showed that the energy consumption does indeed increase strongly along with the income. However, of equal interest is the fact that within each percentile the spread between high and low consumption is almost the same. In other words: there are people who notoriously burden the environment and people who are relatively 'environmentally-sustainable' in each income group. A similar variation, both among and within the income groups, is also perceptible in the United States where it is also evident that ethnic-cultural differences offer a similarly strong explanation for the variation in energy consumption and in the physical elements, such as housing quality and the presence of equipment and, for instance, swimming pools (Lutzenhiser, 1997).

This brings the concept of 'lifestyle' into the picture. To obtain a proper understanding of consumer behaviour and the changeability thereof we should not so much look at the attitudes towards the environment and the availability of environmental information, but rather at the complete lifestyle of consumer groups. This line of reasoning is becoming more popular, but the question is how the concept of 'lifestyle' can be made operational. One way would be to bring spending patterns into the picture as was done in the above. However, this is difficult because the static categories are too inexact, and each Euro for example spent on a holiday can be spent in an environmentally-friendly fashion or not. Another way to look at lifestyles is by comparing materialistic lifestyles with more culturally-oriented lifestyles. Experiencing pleasure from an opera or finding one's amusement in the natural surroundings would therefore be preferred above materialistic consumer behaviour such as going on a shopping spree. However, the question is whether such an opinion would stand on the basis of the factual environmental load. The environmental effect of a consumer's behaviour going on a trip for the weekend to London from Rotterdam, either to do shopping or to go to the opera, will be more or less the same; the same applies to a day of bird spotting on the Waddenzee compared with a day watching the car races at Francorchamps. A general statement as to lifestyles is of little help. The environmental load has nothing to do with a comparison between culture and materialism. What counts in this respect is how much energy and raw material consumption, emissions and residual waste is needed for each form of need satisfaction and the associated mobility.

A more promising consideration is probably to be found in the fact that consumers basically have a limited amount of money to spend (which increases slightly on an annual basis). Seen in this light it can be useful to look at the shifts that occur within this limited budget: is the one activity at the expense of the other activity, which is more environmentally harmful and which less? To do this we need information on the environmental effect per Euro spent. The principle could then be: 'performing the same activity more luxuriously'. This gives us twice the benefit: the cost per unit of need satisfaction increases while, simultaneously, the extra spending makes it impossible to spend alternatively. Moreover, there is then the experience of more quality, and the higher price might be an extra incentive to be more moderate.

The question remains, however, whether this is the right way to approach the problem. Trend analyses and attempts to quantify those analyses, establishing the environmental load per Euro spent, would seem to us to distract from the main task (the task also formulated on the production side; see section 4.2): to improve the efficiency of need satisfaction by a factor of 4 or more. Projects focusing on sustainable technological development are therefore concerned with both production and consumption. The question here could then just as well be: is it possible to achieve a break from the trend in environmental load per unit of consumption or per measure of prosperity or well-being, and in what way and under which societal conditions can it be achieved?

5.4 Addressing overconsumption: eco-sufficiency?

The question now is whether tendencies in consumer behaviour, such as the growth in volume of consumption and 'downturn behaviour' (such as in the laundry behaviour discussed above, or burning the low-energy lamp in the garden at night), threaten to discount the gained efficiency at the level of product and production. Should it become evident that progress in the eco-efficiency of product and production is inadequate, then it will be necessary to change consumption volume by way of a change in lifestyle. In Agenda 21, the concept of 'sustainable consumption' was placed specifically in the context of 'an exceptionally high level of consumption' in western nations with an explicit demand to change value patterns. This is a difficult claim for the authorities to deal with. In a democratic system it is appropriate to assume that the state is not the first and foremost body to impose new social norms and values. The Dutch Government is therefore highly restrained in this respect. This is regrettable because there is a very useful discourse that could be held on overconsumption and the possibilities of sustainable consumer patterns and eco-sufficiency. It would imply that the policy discussion not only looks at making 'more consumption, but with cleaner products' possible, but particularly at the actual volume of consumption.

Initial moves towards such a normative debate are perceptible in civil society. Carley & Spapens in 'Sharing the world' (1998), for instance, explain the constantly growing level of consumption in highly developed western economies as the result of the 'work-and-spend disease'. They see overconsumption specifically as a social problem, characterised by excessive individualism and materialism, whereby individuals are imprisoned in a cyclical, social mechanism: perceiving what others have -> wanting more -> working harder to get it -> earning more -> consuming more, etc. Carley & Spapens demonstrate that this is chiefly a social mechanism (keeping up with the Joneses), fuelled partly by the market's interest in achieving growth and the role of advertising in the mass media. According to the American Merck Survey, a large section of the American population are aware of this mechanism: 88% say that they would change the way they live in order to protect the environment, 82% felt that most Americans buy and consume far more than we need; and 66% would be more satisfied if they could spend more time with family and friends (cited in Carley & Spapens, 1998, pp. 146-147).

An anti-materialistic counter-movement has even emerged in response to this mechanism in the civil societies of several nations such as the USA and the Netherlands. This movement consists of people who wish to disrupt this cycle because they no longer believe in the added value of continually more wealth by way of material goods. Examples are the American Center for a New American Dream which advocates downshifting consumption, the Dutch organisation 'Frugality With Style', and the 'Buy Nothing Days' (see also publications such as Robin & Dominguez, 1992; Van Veen & Van Eeden, 1994; Leiss, 1976).

Such an interpretation of the concept of 'sustainable consumption', focused on a morality of eco-sufficiency, is hardly echoed in today's environmental policy. The accent is on minimising unnecessary wastage of energy and waste and on the route towards improving products, whereby the citizen is addressed as a critical consumer.

5.5 Consumption as the neglected element of environmental policy

While the initiative had been taken earlier, an environmental policy focusing on consumers only came into being in the 1990s in the Netherlands. Back in 1972 the minister responsible for matters of the environment already stated that it was beyond all doubt "that in a society such as ours in the Netherlands, economic growth must be controlled. It is not only a matter of more clean manufacturing, but also a question of being critical in what we consume and leading a prudent life" (Stichting Natuur en Milieu, 1975, p. 93). However, the consumer has only been focused on half-heartedly for many years in environmental policy. While the importance of the consumer's role was always recognised, it was never given much priority. In the 1970s and 1980s citizens were furnished with information and presented with campaigns by environmental and consumer organisations (Ministry of Public Health and Environmental Hygiene, 1972, p. 21; Centrale Raad voor de Milieuhygiëne 1981, p. 30). It is a typically Dutch idea that information on aspects of the environment, especially in terms of making consumers aware and alternative ways of doing things, is much better if it is not given by the government but by civil society organisations – but with the government's financial support. Nevertheless, the government did present some practical, often moralising, biased advice regarding alternative ways to act, relating originally to the use and disposal stage in particular (e.g. when the bottle banks were introduced in the 1970s), and more recently in connection with purchasing behaviour and the use of motorcars ('your car can manage alright for a day without you').

In addition to these communicative instruments, other instruments were also put to modest use in the 1970s and 1980s, statutory requirements or prohibitive rules on products for instance. Also some economic instruments were legally available, like levies (e.g. on petrol) and deposit money, but they were not always applied in practice. We should also mention rules and agreements regarding the use and misuse of advertising in connection with the environment.

From the very beginning, environmental policy has had to grapple with the dilemma as to whether the environmental consequences of increasing levels of consumption should be influenced via the demand side (management focusing directly on the consumer) or the supply side (management focusing on the manufacturers and retailers of the goods consumed). Management via the demand side has the disadvantage that the citizen, or better: the voter, is addressed moralistically, and this does not tend to make the elected policy makers be all that popular. Management via the supply side implies that while product performance can be influenced, the development in consumption volume cannot, and thus the problem could possibly remain unmanageable. It is for this reason that both routes are taken in practice.

The first National Environmental Policy Plan in 1989 marked a paradigm shift in policy focusing on consumers. Households were labelled as a new 'target group' in this plan because of their position as a link in the chain of production – trade – consumption – disposal. This policy document recognised the consumer for the first time as an actor who is able to exert a certain element of power on manufacturers by virtue of his purchasing choices. A threefold strategy was drawn up. Firstly, the consumer was addressed on his purchasing behaviour. In turn, it was made possible for the consumer to react by the second strategy: creating a better supply of information on the environmental consequences. Thirdly, the retail trade was called upon to offer an environmentally friendly range of products. The latter strategy was with a view to stimulating innovations among manufacturers.

For the first time, the objects for several environmental themes for the year 2000 were quantified, such as:

- the total separation of household waste streams;
- a 40% improvement in energy efficiency in 2000 as against the 1985 figures for household electricity consumption;
- a maximum growth of 20% in car mileage as against the estimated growth of 40% over the period 1985-2000;
- a 50% reduction in the use of solvents.

(Ministry of Housing, Spatial Planning and the Environment, 1989, pp. 194-205, 219-223).

The introduction of an ecolabel was also announced, an advertising code of conduct, and the environmental inspection of several important product groups (Ministry of Housing, Spatial Planning and the Environment, 1990, pp. 36-40). At the end of 1993 a separate policy document on Products and the Environment was published. The main objective to which reference was made in this document was to realise a situation in which the market parties, i.e. the manufacturers, the retail trade and consumers, constantly aim towards achieving a reduction in the environmental load of products (Ministry of Housing, Spatial Planning and the Environment, 1993). A great deal of attention in this document was devoted to the further development of a good supply of information about the environmental load of products and the environmental load resulting from their use. To this end, support was given to the development of the so-called life-cycle assessment method and to the development of different forms of information exchange between manufacturers (product registers), between manufacturers and the retail trade, and between trade and industry and consumers. A large part of the document is devoted to stimulating environmentally friendly design and improving production and the retail trade.

Consumption-oriented policy was set in a wider framework in this Product and the Environment document. For instance, it was stated that consumers, who basically are very much aware of the aspects of the environment, must be helped to bridge the gap between attitude and actual (purchase) behaviour by providing them with appropriate facilities, by offering them a good range of products, and by giving them relevant information. The document suggested that this should be achieved mainly on the part of the supply side because the implementation of legislation or the drawing up of agreements is out of the question for this target group. Essential new policies would be: the provision of product information by means of environmental information attached to products, the ecolabel, the setting up of a central environmental information centre, by encouraging the establishment of repair services and recycling shops, and by promoting an environmental purchasing policy among institutional consumers such as the government and other large organisations. By no means all of these plans have become reality to date; the environmental information on products has for the time being been dropped because of the objections raised by the industry. While the growth in activities of recycling shops and repair services has been low, it was not due to policy measures but occurred autonomously. The Dutch Ecolabel (Milieukeur) was introduced in 1992 and, after years of postponement, Milieu Centraal has been operational since 1998 as the planned central information centre on products and can now be contacted by telephone and through the Internet.

Policy focusing on consumption is also pursued by other authorities. At central government level for instance, the more active attitude adopted by the Ministry of Agriculture is relevant in this respect. Since the Plan of Approach for biological agriculture was published in 1996, these products have been stimulated by means of the chain approach. Solutions to problems in the various links of the chain (including production, the wholesale trade, retail trade and marketing) are searched for mutually (Ministry of Agriculture, Nature Management and Fisheries, 1996).

5.6 The effects of policy

Addressing consumers via the mass media

The question is how far this intensified level of attention for the environmental consequences of consumption has started to pay off. We must not forget that many of the policy efforts are aimed at influencing via the supply side: improving the information infrastructure and stimulating the increased supply of products with a better environmental performance. We have already discussed the progress regarding product improvements in section 4.6.

The essence of policy on the demand side is to a large extent geared towards information in the mass media such as the use of short television commercials. Studies in this field show that the direct effect on behaviour is quite low. Nevertheless, there is a lasting wide support for environmental policy and less apocalyptic views on the environment among the Dutch population. The Annual Environmental Behaviour Monitoring study, a longitudinal research on environmental behaviour which has been running for ten years now, shows that in the second half of the 1990s the Dutch population increasingly saw environmental problems as an important issue. Between 1995 and 1999 the percentage stating that the environment was "one of the most important issues" rose from 40 to 47%. The number of persons stating that it was "the most important issue" was between 17 and 20%. Also quite remarkable is the (small) growth in the number of persons stating that environmental issues were the greatest threat to the future: 27% in 1995 and 33% in 1999 (Couvert & Reuling, 2000, pp. 24, 31, 41). Comparable conclusions are also to be found in Hoefnagel et al., (1996) and Ester &Vinken (2000).

What we actually do see is a strong shift in the perception of environmental issues. These issues are rather seen as less of a threat to our health or for our descendants, and a growing number of people feel that "the fuss about the environment is overdone" (Couvert & Reuling, 2000, p. 57). So public opinion tends to shift away from apocalyptic interpretations. We also see a shift in opinions on responsibility. Dutch environmental policy is built on the principle of a shared responsibility of producers, retailers and consumers. Nevertheless, public opinion today tends to see industry as the responsible party; individuals see themselves less as being responsible (Couvert & Reuling, 2000, pp. 50-51).

The perception of one's own responsibility is probably the most significant explanation for the gap between attitude and actual behaviour. Analyses show that this gap is the smallest in the field of waste disposal and energy and water saving, but is particularly wide in the field of nutrition, mobility and purchasing behaviour (see Hoefnagel et al., 1996; Becker et al., 1996, Steg, 1999; National Institute of Public Health and the Environment, 1998; Couvert & Reuling, 1999). In other words, real consumption is kept out of the picture. In this respect we must not forget that these public-oriented studies failed to come to real grips with the problem because, in addition to the gap between attitude and behaviour, there is also a gap between what a person says he does in the answers he gives in surveys and what he actually does in real life. This is because he wishes to give a socially acceptable answer and not always, or not fully, uses a behavioural alternative. For instance: between 1993 and 1996 a total of 80 to 86% of the population said in surveys that they collected their waste paper separately (Couvert & Reuling, 1999, p. 84), while the actual behavioural effect, measured as the share of separately processed waste, grew in that period from 30 to 35% (National Institute of Public Health and the Environment, 1998). In the same period, 86 to 89% claimed to separate their glass waste (Couvert & Reuling, 1999, p. 84), while the actual glass collection in that period grew from 68 to 75% (National Institute of Public Health and the Environment, 1998).

In addition to the acceptance or rejection of one's own responsibility for sustainable consumption, it is quite feasible that the gap between attitude and behaviour can be explained by the quality of the information. Once a search is undertaken for environmental information, the affable consumer is liable to drown in the abundant supply. A list of information material on environmentally friendly behaviour issued by central government, the public utility services, environmental organisations and published in books on the subject, soon adds up to more that 200 behavioural alternatives, varying from 'buy low-energy light-bulbs' and 'do your laundry at the lowest feasible temperature' to 'hang up a nesting box' and 'share your house with other people'. The consumer is unable to see the wood for the trees. The plausibility of the information is at stake here: is it a question of advice in which the environmental merit is evident (e.g. a 10% saving on the annual electricity consumption), or is it a normative or ideological message?

A problem that crops up when establishing a plausible claim regarding the environmental merits of alternative behaviour is that the methods for systematically determining those merits (e.g. life-cycle analyses, environmental measures, eco-indicators, etc.) were only developed in the course of the 1990s. Moreover, these methods are not entirely undisputed as yet. Furthermore, the necessary information is often unavailable as yet and analyses contain assumptions and normative elements which in turn can again be disputed (Vermeulen & De Bok, 2000). This led to a cacophony of environmental assessments and the refutation thereof, especially in the first half of the 1990s.

The effect of social networks and incentives

Approaches other than the mass media are also taken to bridge the gap between attitude and behaviour. The social network approach involves making use of the elements that bind small groups together in order to ensure that awareness and information transfer is more effective. One example in this respect is the ecoteam, an approach developed by Global Action Plan (GAP). These ecoteams, groups of approximately 10 persons (neighbours, friends or co-workers), look at information about the environment and alternative behaviour on the basis of a workbook and discuss what they do themselves, and what they can still do. By recording their own behaviour and by reporting on that behaviour jointly to the organisation, feed back on behaviour is ensured and a certain competitive element is also brought in. Research conducted by Harland & Staats (1995) shows that this approach leads to a further change in behaviour as well as longer-lasting behaviour. One point that should be made here however is that this programme reaches especially those persons that already have an interest in the environment.

Other experiments based on the social network approach also show that it can be considerably more successful than information through the mass media (Weenig, 1994, pp. 205-210). And yet another approach is to make use of emotions, i.e. positive and negative feelings. Regular marketing has also used the emotional element since time immemorial. Exploratory studies show that this approach is able to increase the motivation to actually use the information offered (Midden & Louw, 1994, p. 213).

In addition to management by means of the communicative management model, in consumer policy we also see management by means of instruments conform to the market, such as price incentives. These have been used for quite some time now. Debates have been held on the assumed effects of such instruments. What can we say here on the basis of empirical studies? Vermeend & Van der Vaart (1998, p.45) show that we can expect 1.7 to 2.7 million tonnes of CO_2 will be saved each year from the regulatory energy levy. These expectations are based on model calculations. However, a recent evaluation of the energy levy shows that only half the population is aware of this levy and that no more than 2% of all households take this energy levy into account when using electricity (Staatscourant, 1998). Hence very little can be expected in terms of behaviour.

The effect of product information

As we have already stated, improving the supply of information to consumers plays a pivotal role in policy. What can be said at this point in time as to its effects? First of all we must point out that the Netherlands does not exactly lead the way in the field of eco-labelling. At the introduction of the Dutch Ecolabel in 1992 eco-labelling programmes were in place in a total of 22 countries. In 1999 one or more eco-labelling programmes were available in all the countries of the European Union, as well as in the United States, Canada and Japan. Programmes have also been introduced in various other countries, particularly in the emerging economies in Asia.

Despite this wide diffusion there is little firm evidence as to the effects of eco-labelling. The effect can be measured by the influence on manufacturers (do they change their products and do they apply for an eco-label for new, clean products?) and on consumers (to what extent are they aware of these labels, and do they actually buy these products?). Very few reliable figures are available, especially on the market effects. This is partly because the manufacturers fail to publish these figures. The OECD also notes that it is difficult to make an assessment on the basis of fragmented anecdotal information. A Dutch example in this respect is the large growth in the market share of the first cat litter supplier to market this product with an Ecolabel. The OECD does however conclude that "Producers however continue to apply for and pay for eco-labels, indicating they have some market value" (OECD, 1997, p. 5).

As we can see from Table 5.2 the ecolabel programme was slow to start in the Netherlands. Manufacturers' organisations adopted a reticent attitude regarding their participation in the programme; in some cases the programme was even boycotted. When assessing the effects of an eco-labelling programme not too much value should be attached to these boycotts. They are generally only temporary, as was evident in the past regarding the boycotts of subsidy programmes (see for instance Vermeulen, 1992, pp. 189-195). After some time there is usually one manufacturer or one retailer who sees the opportunity to increase his market share by using eco-labels; these are then followed by others. The OECD concludes that: "Overall, eco-labelling has only been moderately successful with the individual consumer. However, eco-labels may have an important market impact when retailers specify their wish to stock products with eco-labels (for instance ICA in Sweden), or when they become a tool in identifying environmentally preferable products for government procurement and institutional procurement" (OECD, 1997, p. 6).

country	since	name	number of product groups	number of approved products
Germany	1978	Blaue Engel	214	40,000
Japan	1989	Eco Mark	68	3,450
Scandinavia	1991	Nordic Swan	52	2,500
European Union	1992	European seal of approval	14	250
the Netherlands	1992	Milieukeur	27	200
Taiwan	1992	Green Mark	58	800

Table 5.2 *Eco-labelling in some countries: the situation around 1999/2000 (The Global Eco-labelling Network (GEN): http://www.gen.gr.jp/; Stichting Milieukeur, 1999)*

The slow progress of policy focusing on improving environmental information has not been without consequence. The already mentioned Annual Environmental Behaviour Monitoring study has shown that attention given to the environment in purchasing behaviour is on the wane: in 1993 a total of 17% of the Dutch population said to take environmental aspects into account when making purchases, in 1998 this has decreased to 10% (Couvert & Reuling, 1999, p. 84).

Market shares for environmentally approved products

The best way to find out whether policy focusing on consumption is effective is by determining the extent to which there is an increase in the market share of environmentally-friendly products and an actual decrease in the consumption of energy and petrol. Unfortunately, only very little, fragmented information is available on market shares. Better information is available in the field of energy saving. The consumption of electricity is still on the increase (3.4% per annum in the 1990s). Gas consumption figures show a slight decrease as a result of improved insulation methods used in the building trade and the breakthrough of high efficiency boilers for central heating and other forms of sustainable building (Brezet, 1994). In the field of mobility there is no question of a break from the trend for the time being. Ester & Vinken even found that, where they were able to identify a group of green consumers showing consistent beliefs and behaviour in the field of water and energy saving, these green consumers still use their cars as much as other people for their work, shopping and visiting friends (Ester & Vinken, 2000, p. 95). The increase in car mileage was in the region of 2% per annum in the 1990s but now seems to be stabilising, partly as a result of road congestion. Nevertheless, a gradual rise is seen in rail travel (3.5% per annum), while the number of kilometres travelled on public transport is still only a fraction of the kilometres driven by private car (1 to 7-8 km) (National Institute of Public Health and the Environment, 1998).

In section 5.2 we illustrated how the total energy requirement for Dutch household consumption consists for more than two thirds out of activities in the field of housing, heating, nutrition, clothing and caring. Policy focusing on minimising the energy load as a result of consumption will therefore need to be focused to a comparable extent on these everyday household activities and the products used. There are few impressive moves away from the trend to be seen in these fields as yet.

Changing consumer patterns and values?

The slow progress in minimising the environmental load from consumption brings the debate on 'overconsumption' into the picture. Several reports are available on this subject; the general picture is one which shows that we should cherish no high expectations on controlled changes in value patterns in the direction of post-material values. In the Netherlands, Aarts studied the opportunities of diffusion of such post-materialist forms of consumer self-restraint. She expects that "the social trickle-down mechanism, which also explains the diffusion of social values concerning non-smoking behaviour, might enforce 'frugality with style' values." "However," she concludes that "these values remain restricted to a very small group of highly educated / high income people and they are not be likely to be followed as a role model". (Aarts et al., 1995; Aarts, 1999). "A problem here is that the message of 'ever growing levels of consumption having negative effects on the level of (social) well-being' does not ring a bell for most consumers in western societies." The World Values Survey conducted by Inglehart et al., shows us the limits of normative judgements on the consumer behaviour of others. Most people will not perceive high levels of material welfare as conducive to unhappiness. As a matter of fact, high national levels of happiness correlate with levels of economic welfare, even though adherence to post-materialist values is rising in time and in connection with levels of welfare. (Inglehart et al., 1997).

5.7 The changing role pattern among actors

The direct effects of the – still young – consumer-geared policy might well not be in full view as yet, what can be seen is the emergence of new forms of cooperation among the actors involved. The government, the market and NGOs have entered into different forms of cooperation over the past few years. One example being the Stichting Milieukeur, the Dutch ecolabel organisation. The background of this organisation shows that such cooperation does not progress pliantly automatically. The representatives of employers' organisations and the government work together in this organisation. Environmental seals of approval are awarded on the basis of fixed criteria (so-called certification models). A specific procedure must to be gone through in order to ascertain these criteria. Draft criteria are formulated after an exploratory examination. In this stage, the manufacturers, consumers, environmental organisations and scientific experts are able to influence the criteria that is to be drawn up. Despite this cooperation, the organisations of manufacturers called upon their members in the 1990s not to use environmental seals of approval. Apart from that, the ecolabel also has other effects: manufacturers use it when developing their products and are thus able to save themselves valuable time they would otherwise need to spend on establishing essential improvements, even if they do not intend to apply for an ecolabel.

Similarly, a central platform has come into being for the objective of generating reliable environmental information on products and environmental claims made by the manufacturers. This platform, Milieu

Centraal, is run on the same lines in which a procedure of listening to both sides of the argument results in an agreement on environmental claims. This procedure helps to ensure greater uniformity of environmental information and thus avoids the need for businesses to become engaged in discussions with other businesses or environmental organisations on the details of environmental claims.

Such cooperative relationships have also emerged among businesses and environmental organisations. The environmental organisations run product campaigns that bring manufacturers of products with a high environmental performance into the spotlight. Another example is the cooperation between Greenpeace and manufacturers, electricians and energy companies, focusing on the realisation of a lower price for PV cells. This environmental organisation has taken it upon itself to ensure sufficient demand. In eighteen months time there were 5,000 applications for the installation of solar cells – even though more had been hoped for – and Greenpeace turned out to be an unreliable partner (Glasbergen & Groenenberg, 2001).

Another 'relative' success was the appearance of the weekly 'vegetable subscription'. This implies that the producer and consumer enter into a partnership. The producer thus has more certainty as to what he can produce, and the customer is assured of the quality of his food. In 1999 the number of vegetable and fruit subscriptions grew to 45,000. This trend was seen by the government as proof of the professionalisation of environmental marketing (Ministry of Housing, Spatial Planning and the Environment, 1997).

This development marks the new approach taken by the biological agriculture sector over the past few years in which biological food is taken out of the 'alternative' segment. Research shows that a large section of the population, some 30-40%, might be interested in this food, especially because of its image as healthy food (Baggerman & Hack, 1992). For a long time, the market share remained below the poor level of 1%, or even 0.1% for biological meat. The new approach has at least led to a breakthrough in the regular retail trade (vegetable farmers and supermarkets). Sales, which over the last few years had been slowing down and, in comparison with other North West European countries, lagging behind in terms of sales started to grow again in the last few years of the 20th Century. In mid-1999 an increase was noted in the market share of biological dairy produce of 150% compared with 1998 (from 0.6% to 1.5%) (EKO-monitor, 1999, p. 6). In connection with this development, the number of biological agricultural businesses doubled in 5 years (EKO-monitor, 1999, p. 5).

While minor breakthroughs such as these are interesting, they do have their limitations. An initial segment of the market is achieved for the product in question: 2%, 5% or 10%. Occasionally, such as in the foregoing example, a larger segment can be achieved 'in principle'. However, not all consumers will take this road. This also applies in general for non-food products whereby the manufacturers that anticipate the demand for environmentally friendly products are often the manufacturers of the more expensive and higher quality brands. For this group of manufacturers it can probably increase their market share considerably, yet the effect on the entire market is still only small. The question whether these small 'successes' can result in a domino effect and lead to larger 'successes' which have a wider effect on the market is therefore essential; will the rest of the suppliers in the relevant trade (preferably the manufacturers with a large share in the market) react with improvements, for instance by improving their environmental management or possibly with less far-reaching, but gradually progressive product improvements? While these mechanisms have hardly been examined they are still very relevant. After all, an 80% reduction in the environmental load from products for 2% of the market results in a total improvement of only 1.6%. A 20% reduction in the environmental load from products for 98% of the market results in a total improvement of 19.6%. If such a domino effect does occur, then it will be much easier to make the quality requirements at the lower end of the market more strict. Even though in the Netherlands there is a dependency upon developments abroad, similar trends are perceptible there too.

In the Netherlands the government barely anticipates such mechanisms; the strategy is more or less to trust upon the promises of environmentally oriented product development. The government holds the opinion that citizens may only be requested to purchase environmentally friendly goods if the price-quality ratio is comparable with the usual product, otherwise it is being moralising. More is expected from a collective approach in which the government and NGOs (including businesses and trade organisations) interpret the 'sustainability morality' into a range of goals and the associated options for realising them. To achieve this the government is organising consultations with producers, suppliers, retailers and consumers representing the entire life cycle of various relevant area's of consumption, like food, clothing, dwelling, recreation and tourism.

5.8 Reflection: eco-efficient products and eco-sufficient consumption?

In this chapter we have taken a look at how consumption is influenced. Speaking in the words of Rio's Agenda 21, our 'exceptionally high level' of consumption was originally a neglected element of environmental policy but it has now increasingly become the object of policy. If we look at the developments in consumer behaviour, then we can only conclude that realising those goals is for the time being still out of view. Although certain changes have been brought about in consumer behaviour (reuse, gas consumption), policy has still not succeeded in achieving a substantial reduction in the environmental load that results from our consumption habits. We illustrated that the reciprocal dependencies between consumers, manufacturers and the trade call for a good balance between *management,* by way of supply plus the technology applied on the one hand, and *influencing* the demand on the other.

A quite clear trend is emerging in the direction of an emphasis on management via the market parties and increasing the supply of environmentally-friendly products. The consumer is addressed mainly as a rational, calculating consumer. This approach is fully in line with the consensus set out in Chapter 4 on co-responsibility that exists in government circles and the elite of the business community. The primacy to increase the eco-efficiency of production and consumption therefore lies with the manufacturers. It may quite rightly be expected that the manufacturers themselves will be the most capable party to substantially improve the environmental performance of products. However, the question still remains as to the extent that this will lead to a domino effect in the market.

This orientation on the potential of the market is restricted in the sense that it offers absolutely no points of departure for the normative element of the Declaration of Rio. Quite naturally there are two key questions in this respect: do we really need to address overconsumption, and who will be in the position to promote post-materialist values. We can establish that these issues of environmental policy are being ignored for the time being. There are good reasons to state that both state and market are not in a suitable position to carry out this task. Pursuing a public debate on consumer patterns and post-materialist values and lifestyles is a function set aside for civil society par excellence. Following Dutch tradition it can be added to this that the state is able to facilitate that role by supporting the capacity of non-government organisations. An expansion of consumer-related environmental policies in that direction might bring along a good balance between the promotion of eco-efficiency via the market and the promotion of eco-sufficiency via civil society.

References

Aarts, W., J. Goudsblom, K. Schmidt & F. Spier (1995) *Towards a Morality of Modernity,* Amsterdam School for Social Science Research, Amsterdam.

Aarts, W. (1999) *De status van soberheid,* Amsterdam.

Baggerman, T. & M.D. Hack (1992) *Consumentenonderzoek naar biologische producten,* LEI-DLO, Den Haag.

Becker, J.W. et al. (1996) *Publieke opinie en milieu,* Sociaal Cultureel Planbureau, VUGA Uitgeverij, Den Haag.

Brezet, H. (1994) *Van prototype tot standaard, De diffusie van energiebesparende technologie,* Rotterdam.

Carley, M. & Ph. Spapens (1998) *Sharing the World. Sustainable Living and Global Equity in the 21st Century,* St. Martin's Press, New York.

Centrale Raad voor de Milieuhygiëne (1981) *Milieu van jaar tot jaar 1980,* Den Haag.

Couvert, E. & A. Reuling (1999) *Milieugedragsmonitor IX,* NIPO, Amsterdam.

Couvert, E. & A. Reuling (2000) *Milieugedragsmonitor X,* NIPO, Amsterdam.

Eko-monitor (1999), jg. 1999, nrs. 5 en 6.

Ester, P. & H. Vinken (2000) *Sustainability and the Cultural Factor. Results from the Dutch GOES MASS PUBLIC MODULE,* Dutch National Research Programme on Global Air Pollution and Climate Change, Report 410 200 048, Bilthoven.

Eurostat (1998) *Final Household Consumption. Main Results and Detailed Tables 1975-1995,* European Commission, Bruxelles.

Glasbergen, P. & M.C. Groenenberg (2001) 'Environmental Partnerships in Sustainable Energy', *European Environment,* 11, pp. 1-13.

Harland, P. & H.J. Staats (1995) *Het EcoTeam Programma in Nederland. Deelrapport 4: longitudinaal onderzoek naar de effecten van het EcoTeam Programma op milieurelevant gedrag en psychologische achtergronden. Werkgroep Energie- en Milieuonderzoek,* Rijksuniversiteit Leiden.

Hoefnagel, R. et al. (1996) *Milieurelevant consumentengedrag,* Sociaal Cultureel Planbureau, Den Haag.

Inglehart, R. (1997) *Modernization and Postmodernization: Cultural, Economic and Political Change in 43 Societies,* Princeton University Press, Princeton.

Leiss, W. (1976) *The Limits to Satisfaction,* University of Toronto Press, Toronto.

Lutzenhiser, L. (1997) 'Social Structure, Culture and Technology: Modelling the Driving Forces of Energy Household Consumption', in: Stern, P.C. et al., *Environmentally Significant Consumption: Research Directions,* National Academy Press, Washington D.C., pp. 77-91.

Midden, C.J.M. & G. Louw (1994) 'Emotieve factoren in consumentengedrag', in: C.J.M. Midden & G.C. Bartels, *Consument en milieu,* Bohn Stafleu van Loghem, Houten/Zaventem, pp. 212-228.

Ministry of Agriculture, Nature Management and Fisheries (Ministerie van LNV) (1996) *Kracht en Kwaliteit. Plan van Aanpak biologische landbouw,* TK 25127 (1-2), Den Haag.

Ministry of Housing, Spatial Planning and the Environment (Ministerie van VROM) (1989) *Nationaal Milieubeleidsplan. Kiezen of verliezen,* TK 21137(1-2), Den Haag.

Ministry of Housing, Spatial Planning and the Environment (Ministerie van VROM) (1990) *Nationaal Milieubeleidsplan-plus,* TK 21137(20-21), Den Haag.

Ministry of Housing, Spatial Planning and the Environment (Ministerie van VROM) (1993) *Nota Product en Milieu,* TK 23562(1/2), Den Haag.

Ministry of Housing, Spatial Planning and the Environment (Ministerie van VROM) (1997) *Nota Milieu en Economie,* TK 25405(1-2), Den Haag.
Ministry of Public Health and Environmental Hygiene (Ministerie van VoMil) (1972) *Urgentienota Milieuhygiene,* TK 11906(1-2), Den Haag.
National Institute of Public Health and the Environment (RIVM) (1998) *Milieubalans 1998,* Alphen aan den Rijn.
National Institute of Public Health and the Environment (RIVM) (2000) *Achtergronden bij Milieubalans 2000,* http://www.rivm.nl.
OECD (1997) *Eco-labelling: Actual Effects of Selected Programmes,* Paris.
Robin, V. & J. Dominguez (1992) *Your Money or Your Life,* New York.
Slob, A. et al. (1996) *Trendanalyse Consumptie en Milieu,* TNO-STB, Apeldoorn.
Staatscourant (1998), 24 juni.
Steg, E.M. (1999) *Verspilde energie?,* Sociaal Cultureel Planbureau, Cahier 156, Den Haag.
Stichting Milieukeur (1999) *Jaarverslag 1998,* Den Haag.
Stichting Natuur en Milieu (1975) *Natuur en milieu 1972-1974,* Amsterdam.
Van Veen, H. & R van Eeden (1994) *Meer doen met minder,* Aramith Uitgevers, Bloemendaal.
Vermeend, W. & J. van der Vaart (1998) *Green Taxes, the Dutch Model,* Kluwer, Deventer.
Vermeulen, W.J.V. (1992) *De vervuiler betaald. Onderzoek naar de werking van subsidies op vier deelterreinen van het milieubeleid,* Jan van Arkel, Utrecht.
Vermeulen, W.J.V. & C.F.M. de Bok (2000) 'Hobbels op de weg naar een integrale milieubeoordeling van gedragsalternatieven voor huishoudens', *Milieu,* jg.15, nr. 1, pp. 27-39.
Vringer, K. et al. (1997) *Het directe en indirecte energiebeslag van Nederlandse huishoudens in 1995,* Universiteit Utrecht, Utrecht.
Vinger K., et al. (2001) *Nederlandse consumptie en energiegebruik in 2030. Een verkenning op basis van twee lange termijn scenario's,* RIVM, rapport 408129015, Bilthoven.
UN Conference on the Environment and Development (1992) *Verklaring van Rio: Agenda 21,* Distributiecentrum VROM, Den Haag.
Weenig, W.H. (1994) 'Voorlichting in kleine sociale systemen', in: C.J.M. Midden & G.C. Bartels, *Consument en milieu,* Bohn Stafleu van Loghem, Houten/Zaventem, pp. 193-211.

PART 3

ENVIRONMENTAL POLITICS IN A SPATIAL CONTEXT

Chapter 6

THE RENEWAL OF RURAL AREAS

Peter P.J. Driessen, Henri J.M. Goverde and Jan P.M. van Tatenhove

6.1 Introduction

Rural areas have undergone a rapid change over the past few decades, leading to a blurring of what had previously been distinct differences in the Netherlands between rural and urban areas. The Netherlands is a small and densely populated country which in fact has become a single urban field, consisting of both heavily and less heavily urbanised areas which – in the functional sense – are closely linked. While similar developments are also perceptible in other West European countries, they are apparent to a lesser degree than in the Netherlands. This has led to the problems of rural areas taking on a specific character in the Netherlands.

The main function of rural areas in the Netherlands in terms of space is agriculture. Moreover, agribusiness (in this particular case: agricultural businesses and the supply and processing industries) has a huge significance for the Dutch economy. Nevertheless, this is also undergoing change. As a result of the internationalisation of the agricultural economy and the associated crises, the number of agricultural businesses has been dropping steadily for decades now; the total area of agricultural land has also diminished. The agricultural sector is therefore a 'supplier of space' for a variety of other activities, such as the construction of residential areas, nature reserves, recreational areas, industrial estates and infrastructure. These new activities have a radical impact on the actual features of rural areas. Qualities that have been attributed to rural areas since time immemorial, such as 'space' (in the sense of expanse and emptiness), 'peace' (in the sense of quietude and relaxation) and 'beauty' (both in the

P.P.J. Driessen and P. Glasbergen (eds.), Greening Society, 115–135.

aesthetic sense and in the sense of unpolluted), are subjected to an increasing amount of pressure (Driessen, Goverde & Van Tatenhove, 2000, p. 378). The blurring of the boundaries between rural and urban areas has resulted in an increased desire expressed by citizens to enjoy what is regarded as idyllic rural areas. While the agricultural sector becomes more technological, and slowly but surely the rural areas become less rural, the emphasis in commercial activities is being placed more and more on the unspoiled and romantic characteristics of rural areas. However, wherever there is an element of scarcity, prices tend to increase. The recreational sector, for instance, is often seen advertising luxury facilities located in 'natural surroundings' where quietude, relaxation and good health are the keywords. In their advertisements, food producers of exclusive agricultural products stress 'the natural and healthy properties' of their wares by referring to the 'good old days'.

The appearance of the rural areas in the Netherlands is changing: they are no longer simply more 'production space', but are now also turning into 'consumption space'. The aspect of multifunctionality is nowadays taking on more importance. Economic competition, food safety, nature and culture are valued and at the same time stimulated in order to keep rural areas attractive (or to make them attractive) for investments, employment, habitation and leisure activities. Yet we must not forget the quality of the environment. Clean air, unpolluted soil, unadulterated ground and surface water and the careful management of our natural resources are all important conditions for the multifunctional development of rural areas.

Without directly attributing positive significance to this process of change in rural areas, it is nevertheless still a process that can be referred to as 'renewal'. After all, the 'new' *can* be better than the 'old', but this is not always true, certainly not for everyone. It is important to recognise that the changes seen in rural areas are influenced by a variety of societal trends, and that in this respect different value conflicts occur. Within this context an attempt is made to channel the developments in a responsible fashion by means of various forms of government policy, sometimes in conjunction with market parties and NGOs. Simultaneously, the 'good' is constantly under discussion: what is the best path to take? Moreover, the way the developments are managed also undergoes constant change: which developments need to be influenced and what is the most appropriate strategy to achieve that? The renewal of rural areas must therefore be seen as an ongoing process of change in physical manifestations, social activities, discourses, management attempts, and associated societal interactions which are also perceived by the parties concerned as 'undergoing constant change'.

In this chapter we look at the changes in policy on rural areas. In doing so we give particular attention to environmental policy within the context of the more general developments that occur in rural areas. We show how a shift has taken place in the relationship between the three domains of state, market and civil society as a result of societal, economic and political processes. These changes have had consequences for the management method and management strategies used. We also discuss the results that have been achieved and how they can be assessed.

6.2 Changes in agriculture's position in society

In order to understand and assess the changes that take place in the management of rural areas (plus the associated, underlying definition of the problems) we must first of all have an understanding of the societal context of that policy. Considering the main theme of this chapter we shall start by analysing the socioeconomic context by zooming in on the rapidly changing position of agriculture as an economic sector, and on the changes in society's evaluation of this sector.

It has already been pointed out that agriculture is the chief physical spatial function in rural areas and is therefore very important for the quality of the physical environment. There are three developments that bring about the change in agriculture's position in society:

- The liberalisation of world trade for agricultural products.
- The continuing technological, especially biogenetic innovation of food production.
- The animal diseases that occur, and the associated economic, ethical issues and the issues regarding aspects of public health.

The first development, which has a socioeconomic character, can be explained as follows (Goverde, 2000). During the eighth round of the world trade conference, the so-called Uruguay round (1986-1994), the EU was forced to concede to a far-reaching demand made by the USA: that agricultural products must be brought under the terms of the General Agreement on Tariffs and Trade (GATT) and the World Trade Organisation (WTO), which has meanwhile been set up. This means that export and price subsidies on agricultural goods in the EU will to a large extent have to be run down in due course. EU farmers will then have to start producing under world market conditions, i.e. often for a substantially lower price. Furthermore, the EU may no longer protect its internal market from cheaper imports. Implementation of the agreements made in connection with the WTO makes it essential to remodel the Common Agricultural Policy (CAP). The so-called McSharry directive of 1992, aimed at achieving a higher level of market forces in the agricultural sector, flanked by social regulations, must then be pursued. The remodelling of the CAP is therefore shifted from supporting prices to supporting incomes. Given that the Netherlands is one of the world's biggest exporters of agricultural products, these international developments are very important for Dutch agriculture. The production of the Dutch agrocomplex (475,000 man-years, 175,000 of which are in the primary agricultural sector; gross added value: 21 billion euros, 8 billion of which from the primary agricultural sector) is dependent for 75% on exports. It is therefore quite understandable that the Dutch Government sees it as a challenge to secure this contribution to the national economy for the future. Hence a move has been made towards further technologisation, an increase in the level of productivity and a reduction in cost. This also means that the number of farmers will decrease over the next few years, and that production will need to (and indeed can) be carried out on a smaller area of agricultural land.

The two other driving forces mainly influence the image of the agricultural sector as the supplier of food. They are, however, closely related to the internationalisation of the agricultural economy and the increasing level of competition.

As a result of the technologisation of food production, risks have arisen over the past few years regarding the quality of food production. There has been an enormous increase in biotechnological knowledge, resulting for example in genetically modified and irradiated foodstuffs. Irradiation can prolong a product's shelf life because the process kills micro-organisms. Genetic modification leads to fewer diseases in crops and to a higher yield. Nevertheless, some people claim that the technologisation of food production gives rise to environmental risks, the nature and seriousness of which we are still not fully aware. Distinct ethical and normative controversies also arise. Two specific questions are pivotal in this respect: (a) is it sensible to put the health of large groups of people at risk since the consequences of genetic modification cannot be adequately forecast?, and (b) should mankind be allowed to intervene in evolution in this way? After all, biotechnology can also help to reduce environmental risks, for instance the development of crops that can be cultivated without the need to use pesticides.

The welfare of animals and the alleged connection between that welfare and the outbreak of epidemics such as swine fever, foot and mouth disease and BSE is another driving force that has to be dealt with by national and international authorities. The essential, mass slaughter of animals does the image of breeders no good at all. The rapid spread of disease (particularly foot and mouth disease) painfully shows how internationally interwoven the agricultural economy really is, plus all the associated risks. Consumers also make further demands, particularly on ethical grounds, regarding the housing of animals and plants, and an increasing number of requirements are being imposed on the quality of agricultural products from the public health point of view. In general, citizens still expect a certain amount of purity in our green areas in terms of nature, landscape, quietude, odour and beauty. Pollution, stench and noise hinder are not in line with this image (Commissie Wijffels, 2001).

What has always been regarded as a most reliable sector of the economy is nowadays more often being typified as a sector in which the economic, ecological, ethical and health risks have risen to an unacceptable level. In other words, in addition to the pure economic pressure to bring about change, there is also significant pressure on the part of society to run farms differently. Yet the economic and social requirements are incompatible, and this faces the agricultural sector with difficult choices to make.

6.3 Shifts in perception of rural areas

Not only is society's evaluation of the agricultural sector as an economic activity subject to change, there is also a change seen in the way the quality of rural areas as a whole is perceived. Several tendencies are of particular importance in this respect.

First of all, rural areas are today regarded as 'space for expansion' or 'reserve space', that means space for the localisation of a variety of new economic and societal activities. The closer rural areas come to the urban agglomerations, the more pressure they are faced with in terms of space. It is precisely in these areas that a great deal of agricultural land is being converted into land for housing purposes. The Dutch population is still rising steadily. This fact, combined with a heavy tendency towards individualisation (increasingly smaller households) and economic growth (continually higher demands for luxury residences), leads to a heavy demand for new housing projects. It is expected that in the period up to 2030 approximately two million dwellings will need to be built in the Netherlands (Ministry of Housing, Spatial Planning and the Environment, 2001, p. 61). There is also more space needed for new economic activities, for recreational purposes and for the infrastructure (including fast international links). This can only be realised if agricultural land is sacrificed.

Secondly, urban areas are appreciated in another way as well. Life outside the urban area has now become more popular than ever. The rise in prosperity and the increased use of motorcars among citizens has brought the rural areas within reach, both literally and figuratively. The urbanisation process of rural areas is not always a controlled development. It is a process that national government sees as undesirable. Regional and local authorities usually think along different lines. They generally think more in terms of further growth: the financial support of national government, the basis for facilities and the status and salaries of the administrators are coupled to the number of inhabitants. No matter, the revaluation of rural areas does have favourable effects for their economy: a wide range of new economic activity is being established in rural areas, often even in vacated agricultural buildings. There is little difference to be seen in the economic growth in urban as opposed to that in rural areas. Rural areas can therefore no longer be equated to underdeveloped areas, at least not in the Netherlands.

Thirdly, there is a major U-turn perceptible in the ratio of agricultural land to the countryside. For centuries now the agricultural sector has transformed 'wasteland' into 'cultivated land'. The last remnants of natural areas were only just able to be saved from being lost in the 1960s and 1970s. Since the start of the 1990s an attempt has been made to revert the ratios. Partly thanks to society's increased appreciation of the countryside, the government drew up an ambitious plan to develop more nature reserves in the Netherlands (a total of approximately 350,000 ha). It will be mainly agricultural land that will be sacrificed to this end. One important problem is, however, that the agricultural land in many regions has become so expensive (as a result of the demand for space for other societal activities) that policy on the environment is threatening to become almost unaffordable. Meanwhile, farmers have discovered a new source of income from 'countryside activities'. They can count on specific government subsidies from combining agricultural activities with measures to conserve nature or preserve the landscape.

The last trend relates to the changed attitude towards water management in rural areas. The Netherlands not only borders on the North Sea but also on the estuaries of several major (international) rivers. These rivers regularly have to cope with high water and the associated danger of flooding, thus risking the safety of citizens and businesses. In order to reduce the risk of flooding, for centuries now dykes have been built and constantly heightened along the banks of these major rivers. Recently however insights have changed, and the view now is that the constant heightening of the dykes will in due course no longer be effective. In the Netherlands we are also faced with a process of subsidence and, as a result of climate change, a rise in sea level. The new policy strategy is to allocate areas of land that can be used as areas for controlled flooding in periods of high water (Ministry of Transport and Public Works, 2000). This means that a large amount of land will need to be found to receive this water. While these areas will still be able to be used as agricultural land or natural reserves during periods of normal water levels, their allocation as flooding area leads to major restrictions regarding layout and use.

6.4 The dominance of the neo-corporativist decision-making model

For some decades now the development and implementation of policy on rural areas in the Netherlands has been based on the so-called neo-corporativist model. Typical of this model is not only the specific decision-making method, but also the way in which interests are expressed and mediated. The neo-corporativist model can be seen as a special relationship structure between the state, the market and civil society, in which the alliance between state and market is dominant, and civil society is simply side-tracked. To be more specific, in the Dutch situation this model was based on a strong and almost impenetrable bond between the Ministry of Agriculture on the one hand, and the agricultural organisation on the other. This bond was even juridical legitimised by virtue of the configuration of a statutory organisation. The model was focused on one goal only: to reinforce the economic status of the agricultural sector. In consultation with government, the agricultural sector was represented by this statutory organisation, the so-called Board for Agriculture, to which all farmers were tributary (i.e. obliged to pay tax). As a rule, it was always the same representatives that consulted with government. This not only gave stability to the consultations, but was also beneficial for the (usually close) contacts.

The consultations between the two parties were based on consensus; this was also a stimulus for the effective implementation of policy. The government obtained the information it required through this teamwork, a disciplined agricultural organisation and the legitimisation of governmental authority. In exchange, the Board for Agriculture was given information on the government's policy plans, had the opportunity to influence that policy and was given status for the representatives and a monopoly on the representation of interests (Frouws, 1996, p. 32). To ensure that this decision-making model worked, it was essential that the farmers' organisation was a very tightly-knit one. They all shared the same view; a view that was defended at all levels, direct lines to political bodies were also in place. Members of the Board for Agriculture were not infrequently also members of a Municipal Council, a Provincial Council or the Dutch Parliament. This tight-knit agricultural front was apparently very successful in stimulating the economic development of the agricultural sector in the Netherlands. Up to way into the eighties it was impossible for other organisations or authorities that defended other interests in rural areas, and which held different views on the issues and their potential solutions, to penetrate this front. Other organisations, such as the environmentalist and nature conservation movements, were also incapable of building up a countervailing force. Unity of vision and direct lines to political bodies were more difficult to find in these organisations. As a result, the renewal of rural areas was for a long time managed on the basis of a predominant agricultural-economic line of approach. Renewal on the basis of other lines of approach, for instance the development of nature reserves or environmental protection, was initially slow and often came up against opposition. This can be illustrated by two examples: an impairment of nature reserves and the landscape by rural land development projects, and the pollution of the environment due to manure and pesticides.

Rural land development projects

Land development projects are intended to stimulate the development of agriculture in certain areas by way of improving the division of land, water management and the opening up of certain areas. These projects have been carried out on a large scale in the Netherlands since the 1950s, usually in areas where the economic circumstances were exceptionally poor. Land use planning not only contributes towards a rationalisation of management but in fact also to regional development. Slowly but surely these projects came up against strong disapproval. While positive economic effects for agriculture are welcomed, the adverse effects on nature and landscape identity apparently lead to significant opposition (Driessen, 1990). Especially nature conservation societies put up fierce resistance against the large-scale land development projects. Land development may not be at the expense of our natural surroundings, and our valuable cultural landscapes must, according to these organisations, be preserved. However, the reality is quite the reverse. Land development projects radically change the appearance of rural areas. The layout of the land becomes more rational and former, characteristic landscape patterns disappear. On top of this, measures taken in the field of water management (such as lowering the groundwater level in order to make the land accessible for heavy agricultural machinery at an earlier date in the season) evidently have disastrous effects for neighbouring nature reserves, the merits of which are especially thanks to the high level of the groundwater.

Over the years it has become clear that land development could not go on unchanged. The greater importance society started to attach to the nature and the landscape necessitated a change in land use planning. Nevertheless, it took many years before that change materialised. Up to the 1990s, agricultural organisations, in association with the Ministry of Agriculture, succeeded in preventing changes from being made in the policy. Whereas more attention was gradually given to the interests of nature and the landscape, there was still no fundamental change made to the relevant measures or decision-making procedures (which in actual fact were in the hands of the regional agricultural representatives). A 'battle for hectares' gradually developed in several projects (how much land should remain for the agricultural sector, and how much relinquished for nature reserves?), frequently resulting in triumph for the agricultural sector.

Meanwhile, under the influence of developments we shall explain below, times have changed quite radically. One of the main aspects focused on today in land development projects is nature conservation and preservation of the landscape. Representatives of nature conservation organisations are involved in the planning preparations and subsequent implementation. The first steps are even being taken in land development projects to develop new large-scale nature reserves.

Manure and pesticides

The agricultural sector has also had a major influence on the environmental policy on rural areas. Two issues in particular have played a role over the past few years: the excessive production and use of manure as fertiliser and the use of pesticides. Here too has the power of the agricultural sector been an obstacle that has stood in the way of effectuating policy measures.

Since the 1960s, the Netherlands has been home to a growing number of intensive livestock breeding businesses for pigs and poultry in particular. These farms produce an enormous amount of manure which in one way or another has to be spread over the land. The excessive use of manure contributes to soil, groundwater and surface water pollution, as well as to air pollution. While it was observed in the 1970s that the increasing use of manure was leading to adverse environmental effects, it was not until the 1980s that the problem was placed on the political agenda. Various studies have shown that the government – especially the Ministry of Agriculture – had in the meantime done everything in its power to postpone the decision-making process, given that the development of policy on this subject might threaten the continuation of the livestock farming sector (Netherlands Court of Audit, 1990; Termeer, 1993). The problem was played down by the government for a long time and, supported by the agricultural organisations, kept off the political agenda. When in 1984, after alarming studies had been carried out, central government was at last forced to intervene by way of legislation to stabilise the size of pig and poultry farms, the problems had become very serious indeed. The measures that were taken were not, however, in proportion to the seriousness of the problems. Measures such as the enforced reduction in number of livestock were brought forward by several groups in the public debate, but under the pressure of the agricultural organisations they failed to reach the status of the political debate. In other words, while the problem was being tackled it was being tackled in an entirely different way than expected by the (environmentalist) groups, who had hoped to make it a political issue. Policy was tightened up repeatedly over the years that followed, but the farmers in question were always given ample time to adjust to the new environmental standards. Furthermore, the livestock farming sector was apparently particularly inventive in getting round the rules. What is more, it now appears that implementation of the agreed policy displayed many other shortcomings as well.

Policy to minimise the use of pesticides has also shown similar idiosyncrasies for many years: a trivialising of the problems, policy development postponement tactics, fierce controversies regarding the setting of standards, problems in implementing policy and circumventing the regulations by the target group, and as a result the enormous delays in effectuating policy.

6.5 Towards new relationships between state, market and civil society

The tightly-knit agricultural front started to erode strongly in the second half of the 1980s. Frouws (1996, pp. 33-34) mentions two factors that contributed towards this change. The first has to do with the changed demands society made on agriculture as an economic sector. This was dealt with in sections 6.2 and 6.3. The agricultural sector became faced with several quantitative restrictions, mainly in connection with EU market regulations such as milk quotas, the termination of price guarantees for farm produce, and freezing the number of livestock as a result of the manure problem. An increasing number of qualitative requirements were also imposed on the sector. Nutritional and health aspects started to receive more attention. Environmental pollution and impairment of nature and the landscape became political issues. The Ministry of Agriculture was thus forced to drop the exclusive orientation on agricultural production and to pursue a policy on the environment, nature conservation, landscape preservation, recreation, the quality of food and animal welfare. This meant new interlocutors and necessitated a change in management strategies.

The second factor related to changes that had taken place within the farming population itself. A higher level of education, specialisation, as well as an increasing level of technical expertise were the driving forces behind the professionalisation of the farming community. Farmers became emancipated and gradually refrained from allowing themselves be governed by their 'foremen' from within one, tightly-knit organisation. Alternative interest groups were set up, some of which started to adopt a much more radical attitude towards the government. This led to a more fierce battle taking place with that government, particularly on environmental standards and the development of nature reserves, as well as on the strategic development of the agricultural sector as a whole. Instead of an unambiguous presentation of interests, a more differentiated promotion of interests gradually arose and, along with it, discord within the sector.

Certain aspects have had major consequences for the method of management. Since the 1990s, attempts have been made to develop management strategies for the purpose of achieving broad public consensus on the problems surrounding the rural areas and how to solve them. Aided by these new strategies, an attempt was made to ensure that justice was done to the different views and interests; at the same time, it must still be possible to tackle the various authorities and private parties on their responsibilities, and consequently their contribution as well. The development of these new strategies of management is also referred to by the term 'governance'. Governance refers to "sustaining coordination and coherence among a wide variety of actors with different purposes and objectives such as political actors and institutions, corporate interests, civil society and transnational organizations" (Pierre, 2000, pp. 3-4). It relates (as was already stated in Chapter 1) in fact to new patterns of interaction between government and society (Kooiman, 1993). These new patterns are aimed at discovering other ways to deal with societal issues and how to solve them. These 'other ways' (which were also discussed in Chapter 2) relate, inter alia, to the following:

- The widening of responsibility for solving societal problems to include non-governmental parties.
- The recognition that these parties can also hold different views on problems and their solutions.
- The observation that these other parties are able to make a successful contribution towards solving problems.
- The recognition that policy can differ according to time and place, and occasionally *must* differ in this respect.

In the neo-corporativist model, which for many years was the dominant model for policy on rural areas (see section 6.4), there is a tight-knit alliance between the state and market, and civil society is simply side-tracked. In the new way of thinking in terms of the management of the rural areas it is rather a question of collaboration between the representatives of various societal sectors – local governments, businesses and NGOs – in order to develop innovative solutions for the jointly perceived problems (via network management or self-regulation). Rural areas are continually becoming more of an arena in which diversity, plurality and multiformity battle for priority. Various coalitions of public and private actors negotiate on a wide variety of claims for space, conflicting uses, problem definitions and the appropriate options for resolving them (Van Tatenhove, 2000, p. 105). In this arena of varying applications, several attempts are made to bring about an increase in the power of society regarding processes of change that relate to the aspect of sustainability in rural areas. This leads to a shift in the balance of power, changing regulations, discourses and coalitions.

6.6 Different forms of cooperative management in rural areas

Since the 1990s new management initiatives based on these ideas have been undertaken. They relate mainly to forms of region-oriented policy. The four different forms are set out in brief below: the so-called ROM projects, WCL projects, environmental cooperatives and reconstruction projects. We then explain the results that have been achieved with these projects and how they can be assessed.

ROM projects

ROM projects (Spatial Planning and Environmental projects) relate to a special category of problem areas. The integration of spatial planning policy and environmental policy is supported in these areas. ROM projects were initiated in the Netherlands at the end of the 1980s with two goals (Ministry of Housing, Spatial Planning and the Environment, 1988, p. 61):

- To solve problems of a spatial planning and environmental nature, whereby it is essential that a coherent approach is taken. The focus in these projects is on improving the quality of the environment by means of spatial planning policy, and on making spatial developments possible by achieving a clean environment.
- To accelerate implementation of the spatial planning and environmental policy. The problems in these areas are inevitably urgent ones that must be solved, problems which cannot be tackled adequately or quickly enough through the regular channels.

There are six ROM projects, each with a rural character, and all quite sizeable: they relate in fact to two different sorts of areas. In some areas the environmental load is high and the spatial and environmental quality is low. These are sometimes referred to as 'polluted' areas. The prevailing problems in these areas are related to the high concentration of intensive livestock farms, causing a high level of groundwater and surface water pollution, acidification, and the impairment of what is often a characteristic landscape. While the spatial and environmental quality in the other areas is still high, major problems can be expected in due course due to the increasing amount of environmental load from a variety of societal activities (in addition to agriculture, recreation, the approaching urbanisation and traffic). These areas are referred to as 'clean' areas.

The problems dealt with in the ROM projects are by no means new. They are generally situations which were not tackled adequately in the past and have now become out of hand. In virtually all cases is there a question of a stalemate in the regular decision-making process. These stalemate situations must be eliminated by means of network management (see Chapter 2).

One of the main characteristics of these ROM projects is that an attempt is made to mobilise relevant public and private parties to tackle the problem. They are brought into action. By bringing them together for consultation an attempt is made to achieve consensus on the quality of the physical environment and the socioeconomic development options available in the region. All parties then set to work on realising a joint plan. The understanding reached between the parties is then set out in the form of an agreement.

The innovative nature of these ROM projects is not in the use of new policy instruments but in the specific type of organisation: a project-oriented and cooperative approach which is not formalised and which can be concretised in different ways for different regions (Glasbergen & Driessen, 1993). The relevant areas differ vastly in terms of administration, ecology and space. Each region has its own specific possibilities for facing up to the problems. It is presumed that if the right administrative form of policy is chosen, then these possibilities can be used to the best advantage.

WCL projects

WCL stands for 'Valuable Man-made Landscapes'. These projects are not so much related to problem areas (such as the ROM projects described in the foregoing) but to areas worthy of preserving because of their particular characteristics. The areas in question here are those areas that have important natural and landscape values and are of huge significance in terms of culture and history. They also have recreational allure. It is the agricultural sector that mainly determines the landscape in these areas. The functions mentioned are strongly interwoven and there is a certain amount of reciprocal influence and dependency; this can lead to tension, particularly with regard to agriculture, nature and the landscape. Examples being the scaling-up of the agricultural sector, resulting in the possible elimination of small-scale, characteristic elements of the landscape.

All in all, there are eleven such areas designated as WCL in the Netherlands. The objective of policy in these areas, in the abstract sense, is to reduce the tension between agriculture, nature and the landscape, to search for new economic stimuli and to restore the qualities of the countryside, the landscape and the environment. In other words: an attempt is made to reinforce the multifunctionality of the relevant areas while at the same time retaining the aspect of landscape identity. The details of how this should be done can differ from one area to another. To this end, local authorities work together with representatives of the agricultural sector and NGOs in the field of recreation and nature conservation. National government supports the implementation of this policy by making specific sums available.

Environmental cooperatives

Environmental cooperatives are joint ventures on a local or regional scale entered into by farmers who wish to take upon themselves the joint responsibility for the environment, nature conservancy and or landscape preservation while also trying to improve their income by way of innovations (Driessen et al., 1995; Horlings, 1996; Glasbergen, 2000). These cooperatives are in fact a form of self-regulation in which private parties integrate the realisation of the general goals of government policy into their own activities. In this form of management the self-activation of private parties is pivotal. The government only plays a marginal role from the sidelines: establishing the boundaries within which the parties are free to move, and monitoring whether those boundaries are not crossed. The private parties themselves determine how they intend to realise the agreements made with the government and must see that some form of self-regulation is set up within the set boundaries.

While several initiatives were born in the Netherlands some years ago they are quite diverse: from biological farming and agricultural nature conservancy to the marketing of regional products and the search for supplementary income in the recreational sector (e.g. small-scale farm camp sites). Nevertheless, there are several common characteristics to be found in the details of this diversity (Driessen et al., 1995, pp. 95-96):

- The farmers wish to organise their own future in a local or regional form of cooperative collaboration.
- Environmental aspects are incorporated in management in such a way to do justice to the possibilities and limitations of that management.
- When farmers perform better in terms of environmental performance by virtue of doing more, or doing it slightly earlier than others, they demand extra, financial remuneration.

Some of these initiatives were supported by the Ministry of Agriculture for several years. This support implied that the government provided the required scope for policy-making (in the form of dispensation and/or exemption from statutory obligations) in exchange for realising policy objectives in the field of the environment, nature and the landscape.

Reconstruction projects

The fourth innovative form of management is the most recent one and stems from the swine fever outbreak at the end of the 1990s in the Netherlands. Reconstruction projects relate to the restructuring of the pig farming industry. This form of intensive livestock farming, which is concentrated particularly in the southern and eastern parts of the

Netherlands, leads to environmental problems such as acidification, eutrophication, dehydration and nuisance from stench. Furthermore, it threatens the existence of nature reserves. The high livestock density also involves major veterinary risks, i.e. the spread of swine fever. The goal of these reconstruction projects is to achieve a new balance among the different functions (multifunctionality), in which – in addition to a sustainable perspective for agriculture – an attempt is also made to achieve an improvement in the quality of nature, the landscape and the environment. The most important means of doing this is 'zoning'. This implies that pig farming businesses are, where at all possible, concentrated in certain areas far away from nature reserves. Moreover, so-called 'pig-free zones' of 1,000 metres wide are established to act as a buffer against the spread of viral diseases. Pig farms are easily isolated in these 'concentration areas' and thus it becomes possible to prevent disease from spreading. Outside these 'concentration areas' there is not only space for nature, but for the continuation of other agricultural businesses which, subject to strict environmental conditions, contribute to preserving the cultural landscape.

As opposed to the previous three attempts at cooperative management, this form of management makes use of new legal mechanisms (the Concentration Areas Reconstruction Act and the Pig Farming Reconstruction Act) to achieve the goals. Without these legal mechanisms it would be impossible to establish pig-free zones or to relocate businesses. Nevertheless, the preparation and implementation of the plans is done cooperatively, whereby government representatives and representatives of trade and industry and NGOs work in concert. The provincial authorities have the prime responsibility for these reconstruction projects, but separate, so-called 'reconstruction committees' have been set up for each project consisting of the representatives of the local authority, water boards and nature conservancy, agricultural and environmentalist organisations; the advisory members of these committees are representatives of the Ministry of Housing, Spatial Planning and the Environment and the Ministry of Agriculture.

The approach taken in these reconstruction projects reflects a tendency towards a more hierarchical management with a more dominant role for government. Swine fever, and the evidently difficult to overcome nature of such problems as acidification and eutrophication, were reason for central government to take over more control. While the plans are developed at regional level, they are developed within the strict legal and relevant boundaries set by central government. And yet in this respect too can it be established that this is a question of an integral approach being taken towards the problems of rural areas (including the environmental problem), that responsibility is shared by a variety of public and private parties, and that attempts are made to solve the problems cooperatively.

6.7 The significance of the new management efforts

The goal of each of the above four innovative management attempts is to intensify the multifunctional character of rural areas. A specific aim is to find a way of uniting what are often conflicting interests of the agricultural sector, nature, the environment and the recreation sector. This can be done differently for the different regions. One significant feature of the policy efforts is that a generic approach has been abandoned in favour of a regional-specific policy. This creates a tailor-made solution: if the right administrative mould is chosen for the policy, then full advantage can be taken of the opportunities available in the region. The development of policy is inevitably interactive. The authorities responsible for different policy issues work together, often in association with the agricultural sector and NGOs. Nevertheless, there are certain differences between the four innovative forms of management discussed in the foregoing. Whereas the government stands completely on the sidelines as far as the environmental cooperatives are concerned, it is very much present in the case of the reconstruction projects (by virtue of the fact that it draws up the legal and appropriate frameworks). In the case of ROM and WCL projects there is a more or less equality in the relationship between the government, the market parties and NGOs.

Several studies show that positive results can be achieved with these regionally-oriented forms of policy (Wiersinga et al., 1997; Driessen & Groenenberg, 1998; Pleijte et al., 2000). Agreement can apparently be achieved between the various public and private parties on a wide set of measures for a specific area on the basis of 'tailor-made' organisational frameworks. One of the beneficial effects often seen in the projects concerned is a higher level of awareness among the various parties regarding the problems involved, how the problems are perceived by others, and the new options available for tackling the problems.

Yet the significance of these management strategies for the essential multifunctionality of rural areas should not be overestimated. First of all we must recognise that the multifunctionality of rural areas arises automatically by virtue of autonomous forces (i.e. without government policy). As a result of economic developments in the agricultural sector, the number of businesses decreases and along with it the total area of agricultural land. Other functions of rural areas are consequently allocated with – and literally take up – more space: nature reserves, recreational facilities and housing. Multifunctionality is therefore a fact. The question is what kind of multifunctionality must be aimed for, and how can prudent management strategies be used to take advantage of it. There are quite strong differences of opinion on this. It is in fact invariably a question as to whether a choice must be made to preserve the traditional characteristics of rural areas (conservancy) or, conversely, to stimulate the renewal of these areas (modernisation).

A second comment that must be made about the new management attempts is that the policy on rural areas displays a fundamental equivocality (Goverde, 2000, pp. 70-71; Frouws & Van Tatenhove, 1999). Policy, on the one hand, focuses on safeguarding the interests of the Dutch agribusiness by compensating the effects of world market liberalisation as well as possible. This policy is dominated in particular by the EU and national government. Other than agricultural interests play little role in this respect. On the other hand, various initiatives are taken at the regional level with the purpose of in one way or another influencing the multifunctionality of rural areas. In this respect, the authorities no longer play a dominant role but rather enter into agreements with market parties and NGOs. Both developments are more or less incompatible and in fact display the prevalent conflicting value. The dominance of agricultural economic policy in rural areas is still most significant. The regionally oriented policy is in fact only able to try and remedy the most unwelcome effects thereof. There is too little scope for policy for fundamental changes. The regionally oriented policy only has a supporting effect on the mainstream of economic and societal developments. There is a dilemma in this respect too: on the one hand the strong national and supranational management of socioeconomic developments, on the other hand the attempts to generate solutions at regional level. Yet these regional initiatives are only able to have any effect if there is sufficient scope for policy. The cooperative forms of management at the regional level have up to now been unable to produce any major new blueprints for a different sort of agriculture; there are far too few options available. Yet in these regionally oriented projects numerous small-scale initiatives are taken, indicating a paradigmatic change in the economy of rural areas supported by enterprising farmers.

Thirdly, the implementation of regionally oriented policy can be problematic in itself. The success of regionally oriented projects, such as ROM and WCL projects, is still mainly thanks to their process-based nature. In other words, the dialogue and consultations between the authorities, agricultural organisations and environmentalist and nature conservancy organisation are generally productive, often leading to consensus on the desired development direction of a region. Relevant successes do, however, lag behind the process-based successes. Evaluation studies show several obstacles (Goverde et al., 1997; Wiersinga et al., 1997; Driessen & Groenenberg, 1998; Pleijte et al., 2000). As a rule, the objectives of these projects are formulated too loosely, in which the problems and the qualities aimed for (in this case: environmental quality and the organisation of the various functions) are unclear. Furthermore, attention given to monitoring implementation leaves a lot to be desired and thus too little insight is obtained into the government's performance, the performance of the target groups and the developments in the areas in question themselves. The implementation of measures aiming to improve environmental quality or the quality of nature are generally behind measures that focus on developing the (agricultural) economy.

While it is true that the experiment with environmental cooperatives has resulted in a few positive results, the government has still disassociated itself from them. The government lacks the courage to make these cooperative organisations fully responsible for solving societal issues. This means that not only would the problem be taken off their hands, but the actual justification of the approach taken to solve these issues would also be risky. The farmers involved endorse this government attitude simply by claiming scope for policy but failing to accept any collective obligations to perform, coupled to an internal regulatory and control mechanism (Glasbergen, 2000).

Also in this respect is there a dilemma: the dilemma between decisiveness and support (also see Dirven et al., pp. 19-20). Citizens mainly settle with the authorities on the degree to which they are able to solve societal problems, thus demonstrating their determination. The ability of the authorities to solve problems is also limited. There are other actors that play a role in rural areas too and these do not always hold the same views as the government. The different actors involved in rural area development, whether organised or not, each have their own perception of the problems and their own views on the policy measures required. The development of more cooperative forms of management is in this respect no more than logical. The government is therefore reduced to a negotiatory and consultative partner, to being simply one of the players. The intention is to achieve consensus and thus to increase societal support for policy. Yet the ability to solve problems must not only be realised democratically and legitimately, it must also be effective and lawful. Creating support costs a great deal of time, and consensus is not always the same as solving a problem. Furthermore, the government is obliged to follow a variety of procedures and principles of proper administration as laid down by law, and is unable to simply push the responsibilities onto other parties (see also the considerations on this subject in Chapter 2).

6.8 Conclusions

In the Netherlands a major shift has taken place regarding the management of rural area development over the past decade. For many years this management had been based on the neo-corporativist model, a model that was characterised by its tight-knit alliance between the state and the market, or to be more specific: an exclusive relationship of negotiation and consultation between the Ministry of Agriculture and the agricultural organisations. Changes in the social status of the agricultural sector and shifts in society's appreciation of rural areas have resulted in the gradual erosion of this tight-knit relationship. Cooperative forms of management have taken its place, in which an attempt is now being made to do justice to the recognised different views and interests regarding the problems of rural areas. This involves new forms of collaboration between the authorities' representatives, the representatives of the agricultural sector and NGOs at regional level. At the same time, a wider approach is being taken to the problems of rural areas. While the renewal of rural areas has been managed for many years mainly from an agricultural-economic angle, an attempt is now being made to shape multifunctional rural areas, taking into consideration issues of environmental protection, nature conservancy and socioeconomic development collectively.

It is often claimed that it was not so much a question of *deciding to* apply these new forms of management, but that it was *inevitable*. Society is developing further in a network fashion and this calls for other methods to solve the problems. The new methods of management are seen as essential everywhere and are therefore latched onto positively. More initiatives are being taken to find new alliances between the state, the market and civil society at the national, regional and local level. Openness, generating support and achieving consensus are the keywords in this respect.

However, the expectations regarding the new forms of management should not be overestimated, we argued this in the previous section. Agriculture is still a very dominant factor of rural life, although its social significance is undergoing change. Radical changes in rural areas cannot be achieved without a radical restructuring of that agriculture itself. The economic, ecological, ethical and health risks the sector itself allows to arise must be brought back to an acceptable level. As yet, regionally oriented projects, such as ROM and WCL projects, environmental cooperative projects and reconstruction projects, are unable to contribute adequately to this given that agricultural development is steered chiefly by national and European policy. It is increasingly recognised that rural policy can and must differ according to place and time. At the same time, the options for doing this are apparently far too limited because the main actor in rural areas, the agricultural sector, is addressed in particular by a very uniform policy.

The new forms of management discussed in the foregoing can therefore be all the more successful if more scope is offered in policy for decentralisation and differentiation. This means a greater role in terms of content for the individual regions in the shaping of policy. Changes are gradually taking place in this direction. In the framework of rural area development we now see complex relationships developing between Europe and national and regional administrations. The EU is currently not only addressing the regions directly or through the Member State in question, via the INTERREG and LEADER projects for instance, regional authorities are becoming increasingly aware of how to circumvent their national governments and establish contact with Brussels direct. Simultaneously, we also see various initiatives being exchanged between regions (Van Tatenhove, 2001).

Regional diversity in development not only makes rural areas more attractive for investments, housing and leisure activities, it can also make the agricultural sector more competitive in a different way. The motto to be adopted in this respect should be 'more quality, more diversity and a greater interweave with nature and the landscape'.

References

Commissie Wijffels (2001) *Toekomst voor de veehouderij; agenda voor een herontwerp van de sector,* Den Haag.

Dirven, J.M.C., I.Th.M. Snellen, P. Rademaker & J.A. Schild (1998) *Stuur of overstuur; over bestuurlijke wisselwerkingen tussen overheid en samenleving,* Elsevier bedrijfsinformatie bv, Den Haag.

Driessen, P.P.J. (1990) *Landinrichting gewogen; de plaats van milieu-, natuur- en landschapsbelangen in het landinrichtingsbeleid,* Kerckebosch, Zeist.

Driessen, P.P.J., P. Glasbergen, P.P.P. Huigen & F. Hijmans van den Bergh (1995) *Vernieuwing van het landelijk gebied; een verkenning van strategieën voor een gebiedsgerichte aanpak,* VUGA Uitgeverij, Den Haag.

Driessen, P.P.J. & M.C. Groenenberg (1998) *Monitoring van gebiedsgericht milieubeleid; een analyse op provinciaal niveau,* RIVM, rapport nr. 251701034, Bilthoven.

Driessen, P.P.J., H.J.M. Goverde & J.P.M. van Tatenhove (2000) 'Vernieuwing van het platteland', in: P.P.J. Driessen & P. Glasbergen (red.) *Milieu, samenleving en beleid,* Elsevier bedrijfsinformatie bv, Den Haag, pp. 377-398.

Frouws, J. (1996) 'Politieke modernisering van het groene front?', *Sociologische Gids,* nr. 1, pp. 30-45.

Frouws, J. & J. van Tatenhove (1999) *Regional Development and the Innovation of Governance,* Working Group 4, EU-Cost A12, Vienna.

Glasbergen, P. & P.P.J. Driessen (1993) *Innovatie in het gebiedsgericht beleid; analyse en beoordeling van het ROM-gebiedenbeleid,* Sdu Uitgeverij, Den Haag.

Glasbergen, P. (2000) 'The Environmental Cooperative: Self-Governance in Sustainable Rural Development', *Journal of Environment & Development,* vol. 9, no. 3, pp. 240-259.

Goverde, H.J.M. et al. (1997) *Bestuurlijke evaluatie Strategische Groenprojecten natuurontwikkeling,* Wageningen.

Goverde, H.J.M. (2000) 'Waardeconflicten over rurale ontwikkeling', in: N. Nelissen, H. Goverde & N. van Gestel, *Bestuurlijk vermogen; analyse en beoordeling van nieuwe vormen van besturen,* Uitgeverij Coutinho, Bussum, pp. 43-76.

Horlings, I. (1996) *Duurzaam boeren met beleid; innovatiegroepen in de Nederlandse landbouw,* Nijmegen.

Kooiman, J. (ed.) (1993) *Modern Governance. New Goverment-Society Interactions,* Sage Publications, London.

Ministry of Transport and Public Works (Ministerie van V&W) (2000) *Anders omgaan met water; waterbeleid in de 21e eeuw,* Den Haag.

Ministry of Housing, Spatial Planning and the Environment (Ministerie van VROM) (1988) *Vierde nota over de ruimtelijke ordening,* Den Haag.

Ministry of Housing, Spartial Planning and the Environment (Ministerie van VROM) (2001) *Vijfde nota over de ruimtelijke ordening; Ruimte maken, ruimte delen,* Den Haag.

Netherlands Court of Audit (Algemene Rekenkamer) (1990) *Decemberverslag 1990,* TK 21955/1-2, Den Haag.

Pierre, J. (ed.) (2000) *Debating Governance. Authority, Steering and Democracy,* Oxford University Press, Oxford.

Pleijte, M. et al. (2000) *WCL's ingekleurd; monitoring en evaluatie van het beleid voor waardevolle cultuurlandschappen,* Alterra, Wageningen.

Termeer, C.J.A.M. (1993) *Dynamiek en inertie rondom mestbeleid; een studie naar veranderingsprocessen in het varkenshouderijnetwerk,* VUGA Uitgeverij, Den Haag.

Van Tatenhove, J. (1996) 'De regio als beleidsarena', *Sociologische Gids,* no. 1, pp. 46-59.

Van Tatenhove, J. (2001) 'De meervoudige groene ruimte', in: H. de Haan et al. (red.) *Het precaire evenwicht tussen distantie en betrokkenheid,* Wageningen, pp. 98-113.

Wiersinga, W.A. et al. (1997) *Evaluatie ROM-aanpak,* Ministerie van VROM, Den Haag.

Chapter 7

POLICY INNOVATIONS IN THE URBAN CONTEXT

Jochem F.M. van der Waals and Pieter Glasbergen

7.1 Introduction

Urban areas are the cause of two sorts of environmental problems. The first sort is seen in the cities themselves, with noise pollution, local air pollution, external safety risks, and soil pollution. There are also those environmental problems that cities push onto other areas and the higher levels in terms of scale. We are concerned here with the emissions of pollutants to air and water, the production of urban waste, and the demands of production and consumer processes on water, energy and other resources. The emphasis in this chapter is on the environmental problems inside the urban areas themselves and less on the environmental problems they cause for their surroundings.

The paradigm shift in Dutch environmental policy has also had an effect on the approach taken to urban environmental issues. In this chapter we discuss the policy innovations in the urban context and the backgrounds of these innovations. There are three different kinds of innovation. Firstly, the decentralisation of the decision-making process on certain elements of environmental policy to local authorities. Secondly, environmental policy and spatial planning are to an increasing extent becoming integrated, whereby the environment is no longer seen as the automatic limiting condition for spatial planning but is rather weighed against the economic and societal objectives. Thirdly, there is an increasing level of participation among citizens, NGOs and businesses in urban environmental policy. These innovations have had consequences for such policy topics as noise pollution, soil cleaning, sustainable building and Local Agenda 21. More

P.P.J. Driessen and P. Glasbergen (eds.), Greening Society, 137–160.

integrated planning methods have also been developed to tackle the environmental issues in urban areas.

In the following we present a brief outline of the societal and spatial determinants of urban environmental problems. We then discuss the policy innovations and the associated opportunities and limitations. Finally, we move on to discuss the question how these innovations should be assessed.

7.2 The societal backgrounds of urban environmental problems

Urban environmental issues originate from a variety of societal developments. One initial, significant development is the increase in size of the urban area itself. Between 1970 and 1996 the total urban area in the Netherlands increased by 64% (Ministry of Housing, Spatial Planning and the Environment, 2000a). Especially the growth in number of households was the reason for this. The use of urban space per head of the population has increased because of the decrease in average housing occupancy. Urbanisation is seen mainly in the urban areas of the western, central and southern parts of the Netherlands. This total area covers the Randstad, a circle of cities situated around the so-called Green Heart of the Netherlands, and the cities to the east and south of the Randstad in the Provinces of Gelderland and Brabant. Particularly those municipalities in the vicinity of the large cities have grown as a result of migration from the cities, especially in the 1970s (National Institute of Public Health and the Environment, 1998, p. 29). The employment percentage has also risen more strongly in the neighbourhood of the cities themselves as a result of the lack of space, urban congestion and the high cost of business accommodation for space-intensive industries (Van der Vegt & Manshanden, 1996). More in general, the level of urbanisation is gradually expanding and, as was already indicated in Chapter 6, this is difficult to keep in check in rural areas. Because businesses and households are gradually operating on a larger scale, the differences between urban and rural areas are becoming less pronounced. In this connection we speak of urban fields.

To a certain extent we can bring the impact of urbanisation in terms of the use of space into perspective. In 1996 about 11% of the total area of the Netherlands was used for urban functions (Ministry of Housing, Spatial Planning and the Environment, 2000a). The influence of urbanisation on rural areas must be seen more in terms of the indirect effects on other areas. For instance, urbanisation can lead to a fragmentation of animal biotopes and thus to the risk of a drop in the size of species populations. Also the construction of the infrastructure and the associated mobility and recreation around the urban areas can lead to adverse ecological effects.

The second major development is the increase in level of mobility. This is to a large extent the result of demographic and socio-cultural factors.

Between 1970 and 2000 the Dutch population grew from 13 to 15.9 million inhabitants. In that same period, the number of households increased even stronger; by about 65% to approximately 6.7 million (National Institute of Public Health and the Environment, 1998). Especially the number of one and two-person households has grown. The increase in number of households, particularly in the age group of 25 to 50-year-olds, has had a considerable effect on the rise in personal mobility. The growth in prosperity and rise in number of jobs has also contributed to the higher level of car ownership and car use. The growing demand for mobility had been stimulated by an increase in the number of new roads and an improved supply of public transport. Heavy goods traffic has also undergone substantial growth (Consultancy Service for Traffic and Transport, 1997).

Road transport leads to diverse environmental problems. This form of transport is the main source of noise pollution, followed up by noisy neighbours and air traffic, and to a lesser extent industry and rail transport. Noise pollution is relatively often present in urban areas (National Institute of Public Health and the Environment, 1998, p. 124). Apart from that, the current increase in amount of traffic is seen mainly outside the urban areas and on the main roads (National Institute of Public Health and the Environment, 2000, pp. 123-124).

Road traffic is also one of the main reasons for the relatively high level of air pollutant concentrations in Dutch cities. Increased concentrations of hazardous substances, such as carbon monoxide, nitrogen oxides and benzopyrene is found alongside heavily used roads in particular. Nevertheless, despite the increase in mobility, the air quality in the cities has been improved considerably thanks to the introduction of catalytic converters, the use of cleaner engines and fuels and traffic routing (National Institute of Public Health and the Environment, 1998, p. 131).

The third factor is the development of industrial production (see also Chapter 4). Industrial areas are often situated inside or in the vicinity of urban areas and consequently impose an environmental load on the neighbourhood. The main emissions in this respect are SO_2 (power plants, refineries), NO_x (power plants, refineries, (dust) particles (base metal, chemical industry), and VOS (chemical industry, refineries). While these emissions have dropped considerably since 1990 it does not eliminate the fact that major problems still arise locally, such as in the major industrial areas around Rotterdam and Amsterdam (National Institute of Public Health and the Environment, 2000, p. 125). At certain locations in the city do we see external safety risks attached to the use, storage and transport of hazardous substances, as is the case in the chemical industry, and in railway yards. The strong economic growth in the Netherlands in the 1990s took place to a large extent in the less-polluting services sector rather than in industry. However, it should be pointed out in this respect that the growth in the services sector has resulted in an increase in mobility, also in urban areas.

Finally, our inheritance from the past is also the cause of urban environmental problems. Plans for house building, business premises and urban renewal are occasionally postponed because large areas of land are evidently polluted by businesses that stood there in the past. Houses have been built in the centre of old towns on a metres high 'urban layer' formed by centuries of dumping municipal waste: domestic waste, the remains of old houses, plus the waste from small industries such as tanneries and small paint factories. The sites of former gas factories located outside the old city centres are also seriously polluted. In a wider sense, the historically grown urban structure can also be seen as one of the causes of environmental problems. Aspects such as the proximity of industry and residential areas and the narrow traffic routes are often difficult to change.

The different environmental problems in urban areas are distributed unequally in terms of space. They therefore also have a social dimension in the sense that they cause a great deal of inconvenience for certain social groups. For instance: studies conducted in Tilburg and Rotterdam show a clear concentration of environmental load in the city centres and the surrounding districts. These have to put up with heavy traffic, the proportion of multi-story houses is high here, and accessibility to green areas is poor. People on low incomes and with a low level of education are over-represented in these areas (National Institute of Public Health and the Environment, 1997, pp. 113-119).

7.3 The spatial dimension of urban environmental problems

The method of approach taken to deal with urban environmental problems is made more complex due to the spatial factors involved. For instance, the density and the actual combination of functions in the city contributes towards noise pollution. Local air pollution and odour is intensified by the spatial concentration of activities in urban areas. As a rule, the bigger the city, the higher the level of concentration of air pollutants. Conversely, the concentration of activities is favourable for limiting those environmental problems that cities push onto other areas. The grouping together of urban areas is often seen as a way to limit the increase in mobility, promote the use of bicycles and public transport and to protect and preserve open areas and nature reserves. These presumed environmental benefits are therefore one of the arguments in favour of the concept of 'the compact city'. Because compact urbanisation has both benefits and disadvantages for the environment and spatial planning, we speak of the 'dilemma of the compact city'.

Nevertheless, the environmental benefits of the compact city are not as high as sometimes assumed. One of the main considerations that led to the compact city policy is the anticipated limitation of motorised traffic resulting from the short distance between the areas where people live and where they work, as well as the higher level of support for public transport. Various studies show that a high level of building density in the vicinity of cities and a combination of functions can achieve a certain reduction in the amount of traffic, but that this reduction is still limited. In general, individual and household characteristics such as income, education, age and position on the labour market reflect more cohesion with mobility behaviour than the spatial environment in which one lives. Economic and demographic developments such as those discussed in section 7.2 consequently have more of an effect on mobility than the spatial structure of urban areas (Verroen et al., 1995; Anderson et al., 1996; Van der Waals, 2000).

A comment can also be made regarding the effect of urbanisation on the countryside. According to the National Institute of Public Health and the Environment (1997), the main cause of the reduction in number of species since 1900 was the combined effect of eutrophication, acidification and dehydration. The effects of urbanisation on the countryside are small in comparison with the influence the agricultural sector has on the quantity and quality of nature. Too great an emphasis on compact urbanisation also implies a reduction in the ecological quality of urban greenery. Nevertheless, compact urbanisation offers better opportunities for the development of large-scale, unbroken nature reserves (Farjon et al., 1997).

Because of the connection between, and the dilemmas of the various environmental aspects, not only have new policy initiatives arisen for sectoral environmental problems, but we also see initiatives focusing on integration and the consideration of various environmental aspects. In the following we discuss how the policy on urban environmental problems has developed. We start with two 'classic' environmental issues: noise pollution and soil cleaning.

7.4 Noise pollution

Noise abatement policy is the perfect example of a field of policy which is built on the traditional, hierarchical management-based paradigm, underpinned by scientific knowledge (see Chapter 1). An essential element of noise abatement policy is that limiting conditions must be imposed on spatial development. To minimise the effects of noise pollution, 'noise zones' are first established which subsequently function as a buffer between the actual sources of noise, such as industry, road traffic and rail transport, and 'noise sensitive locations' such as homes, schools and hospitals. These zones are established on the basis of national standards. The Noise Abatement Act contains an extensive system of regulations to combat noise pollution. For instance, it establishes limits on the different sources of noise; a base limit (the preferred limit) and an upper limit (the legally allowed maximum noise level). Policy on noise abatement has led to a reduction in the number of people experiencing discomfort from noise by imposing stricter quality requirements on newly built homes (noise insulation), the introduction of speed limits and mitigating measures such as the erection of noise protection walls (National Institute of Public Health and the Environment, 1998, p. 128).

Local authorities are the main bodies responsible for implementing these nationally-established limit values. They have expressed a great deal of criticism on the Noise Abatement Act because of its high level of detail, high level of complexity and the high costs involved in its implementation. Another bottleneck experienced is that the uniform noise standards allow too little scope for taking local aspects into consideration. In certain cases, strict observance of the noise standards apparently gives rise to too many restrictions on spatial development, making it difficult to accomplish the aims of realising compact urbanisation.

National government responded to this criticism by starting up a project that focuses on bringing the noise abatement policy more up to date (as well as the underlying legal framework). The project was implemented in collaboration with provincial and local authorities and must take effect in 2002. The review of the Noise Abatement Act is in line with a wider trend towards minimising the complexity and level of detail of Dutch legislation.

Pivotal in the modernisation of the noise abatement act is a decentralisation of the decision-making process. Local authorities are given greater powers to pursue their own policy on noise abatement. In the new system, central government is still responsible for formulating the statutory guidance values for noise levels measured on the outside walls of dwellings and other noise sensitive locations, but providing they are able to motivate their reasons, local authorities may deviate from these values. These guide values also act as the limit values for those local or provincial authorities that do not have their own noise abatement policy and associated limit values.

The idea behind this decentralisation is to provide more scope for customised solutions in which attention can be devoted to the noise quality to be achieved in the different areas, and the discomfort experienced by the inhabitants and users of an area. Societal and economic interests can also be taken into account. Local authorities have more opportunity to condense functions within the framework of policy on the compact city. This must achieve a better integration of environmental policy and spatial planning. Moreover, the underlying idea is that it is much better for local authorities to be tackled on their noise abatement policy by the citizens themselves. The quality of the policy must not so much be safeguarded by higher authorities, but rather by an open process of decision making. Local authority noise abatement policy must be realised after consultation with national government, the provincial authorities, businesses and NGOs, and made definitive by the municipal council.

7.5 Soil cleaning

The second field of policy recently reviewed is policy on soil cleaning. Soil pollution in urban areas is problematic because the poor environmental quality imposes limitations on functions such as housing and recreation on polluted locations. As also applies with regard to noise pollution, the original basis for this policy was a model in which environmental policy dictates the standards that accommodate the limiting conditions for spatial planning. A detailed legal framework has also been developed which binds the municipal authority to the policy pursued by both national government and the provincial authorities. In the Soil Protection Act the provincial authorities and the four main municipal authorities (Amsterdam, Rotterdam, The Hague and Utrecht) are designated as the 'appropriate authorities' and consequently play the pivotal role in assessing and implementing soil clean-up plans.

The original aim was to realise soil 'multifunctionality', i.e. restoration of the use functions of the soil for man, plant and animal. At a later date it was established that, subject to certain conditions, instead of removing polluted soil the decision could be taken to 'package' the pollution itself (isolate, control and inspect). The Soil Protection Act was also based on the principle of 'the polluter pays': the person that causes the pollution (or if this is out of the question, the user of the land) must pay for the costs of soil cleaning operations.

It has since become apparent that the current set of policy instruments is inadequate for tackling these problems. The magnitude of polluted areas is evidently far greater than originally assumed, and the budget is insufficient to carry out all the clean-up plans. In practice, it appears virtually impossible to recoup the clean-up costs from the polluter (Interdepartmental Study Group on Soil Cleaning, 1997).

A process of policy innovation was started up in 1997 as a joint effort between national government and the provincial and municipal authorities. In the new approach, the principle of multifunctionality was abandoned and replaced with 'function-based soil cleaning'. This means that the soil is cleaned up to the level of quality required for the specific use in question. The principle that the soil must be suitable for any function was therefore dropped. This meant that local circumstances and spatial and economic developments could be taken into account. Like noise abatement policy, here too have the environmental requirements been made more flexible. The soil's environmental quality is no longer the factor that determines the options for the functions to be realised; spatial development is now the determining factor for the required soil quality. To make function-based soil cleaning possible, the regulations are being adjusted and simplified. Various tasks are also being decentralised. Thirty large municipal authorities are being given the role of 'appropriate authority' unless they express the wish not to be designated with such a role. While there is still the opportunity to work on the principle of 'customised' solutions, general rules are drawn up for a standard method of approach to certain types of cases per region (Regiegroep Bever, 2001).

The idea behind this new approach is that clean-up operations can be carried out cheaper, quicker and with a higher contribution from the market parties. A so-called 'system of participation' is to be introduced in which soil cleaning can be jointly financed by government and market parties such as land owners, project developers, land administrators and future users (Regiegroep Bever, 2001). The financing of soil-cleaning projects was brought into the Investment Budget for Urban Renewal (the so-called ISV), introduced in 2000. This ISV is a combination of several major flows of funds in which, in addition to resources for soil clean-up operations, subsidies for urban renewal and restructuring, large new housing development locations, improvement of industrial estates and the insulation of dwellings from noise pollution are integrated. The ISV also reflects the wish to provide local authorities with more scope for their own policy making, and to improve the integration of policy. Financing is granted on the basis of a municipal development programme in which the objectives must be defined on the basis of twelve broadly formulated fields of achievement. The ISV sums involved are undertaken for a period of five years.

7.6 The decentralisation and expansion of flexibility: an interim account

We see quite clear similarities in the methods of approach taken on both noise pollution and soil cleaning. The government traditionally manages both of these policy themes in a hierarchical fashion, establishing and implementing standards on the basis of scientific knowledge as to the environmental effects for the entire nation. This approach comes up against problems of implementation: the societal consequences of strictly applying the standards (restrictions on house-building projects, a stagnation in spatial development due to soil clean-up operations) are – at least in a number of cases – found to be unacceptable. Solutions are sought in decentralising the decision-making process and in creating scope for policy-making at local level. Scope is created for making concessions regarding the desired environmental quality, thus making spatial development possible. In certain cases, the less strict application of environmental standards are justified in order to achieve a good quality of the physical environment in its totality. This is referred to by De Roo (1999) as the 'balance method of approach', an approach characterised by its wider, jointly established objectives in terms of both the environmental and the spatial quality to be achieved over a longer period of time. ROM projects (see Chapter 6) are in line with this trend.

However, the question now is how this trend towards decentralisation and flexibility should be assessed. One advantage is that the policy's practicability is wider. Moreover it creates broader scope for taking the local differences into account; thus the aspect of noise can be dealt with more flexibly for the inner city than for the suburbs. Furthermore, it also makes it possible to do much more in terms of other matters that are important for the city; bringing in new commercial activities for example.

One of the drawbacks of the balance method of approach – if the decision-making process is left up to the municipal authorities – is that the environmental quality might possibly be made subordinate to spatial and economic development. After all, municipal authorities are concerned far more with local economic interests than are the higher authorities.

The second objection is that the policy is very much dependent on the knowledge available in the municipal authority, its capacity and its ability to develop strategic policy. There is thus always the risk of municipal authorities being forced to rediscover the wheel. According to Hut & De Roo (1999) municipal authorities are hardly equipped for expanding their package of tasks regarding soil cleaning operations. Strategic policy on soil cleaning has only to a limited extent been developed to date.

Another risk is that, because of the lack of an unambiguous standard, it is difficult to tackle municipal authorities effectively on their policy. The question is how far decentralisation really does go hand in hand with a significant amount of participation among citizens and other stakeholders.

The tendency to take a more coherent look at the different aspects of the physical environment gives rise to the question how environmental aspects should be taken into consideration, one against the other. In the next two sections we look at two ways in which an attempt has been made to methodically shape this integral aspect of study: by using the so-called environmental assessment method and by adopting the City and the Environment approach.

7.7 The method of environmental assessment

Over the past few years, several municipal authorities have developed methods to make the 'dilemma of the compact city' more understandable and more manageable, and to improve the level of coordination between spatial planning and environmental policy. These so-called 'environmental assessment methods' are intended for the purpose of assessing the environmental impact of plans for spatial developments in the city in advance, and/or establishing the limiting conditions for spatial developments on the basis of their environmental significance. Some environmental assessment methods relate to strategic municipal policy, others to its implementation. The methods also differ in terms of purpose, level of scale, degree of detail and the point at which they can be used in the planning process. The assessment methods can be divided into 'normative zoning methods' and 'balance-oriented methods' (Humblet & De Roo, 1995).

Normative zoning methods are based on scientifically established environmental standards and are focused on the field of tension between two functions (e.g. the dwelling function and the function of work). On the basis of the actual source of tension a comparison is made of the encircling field and the associated environmental load, the outcome is then used to establish environmental zones. These methods are often characterised as 'investigatory'. A distinction can be made between 'inward zoning', in which the required distance from the environmental sensitive activity (dwelling) to the industrial activity is established, and 'outward zoning', in which zones are established around functions that have an impact on the environment and which involve restrictions on the use of the space located inside these zones.

One example of a normative method is the 'Voorlopige Systematiek voor Integrale Milieuzonering' *(Preliminary System for Integral Environmental Zoning)* (VS-IMZ). The Ministry of Housing, Spatial Planning and the Environment has been working on the development of this method since 1989 for the environmental quality around industrial complexes. The intention was to convert the method into a statutory regulation. The method included classification and the weighing of standards for industrial noise, odour, external safety and hazardous substances, translated into six integral environmental zones. Of these, only the noise standards have legal status.

The state of affairs regarding the VS-IMZ illustrates how difficult it can be to establish environmental zones and translate them into consequences for the spatial organisation of areas. While the method has been used in 11 experimental projects, it was not incorporated in policy because of the large number of problems involved. For instance, it was difficult to establish the accumulation of environmental loads in a scientifically responsible manner. Moreover, it turned out to be far more difficult to specify zones on odour and external safety than on noise because they are not limited to a single industrial site and may even be regional by nature. The legal practicability was also restricted given that some of the environmental standards had no legal grounds. The most problematic aspect was probably that in certain cases substantial environmental zones emerged which would lead to factories having to be shut down, residential areas demolished, and the blockage of new spatial developments (Borst et al., 1995).

Balance-oriented methods look at environmental quality in a wider urban area and are based on environmental quality objectives for the area as a whole. These methods are more general by nature, give consideration to several aspects and interests, and consequently their function is chiefly strategic. A well-known balance-oriented method is the so-called 'city dome' (see Figure 7.1). This was introduced in 1994 by the Amsterdam Council to ensure that the considerations in the field of environmental policy and spatial planning were more overseeable. An imaginary dome was placed over the city, reflecting the actual environmental load in the city.

The goal of this policy is to minimise the environmental load. The 'dome' is divided into 'smaller domes'. The idea being that not all areas of

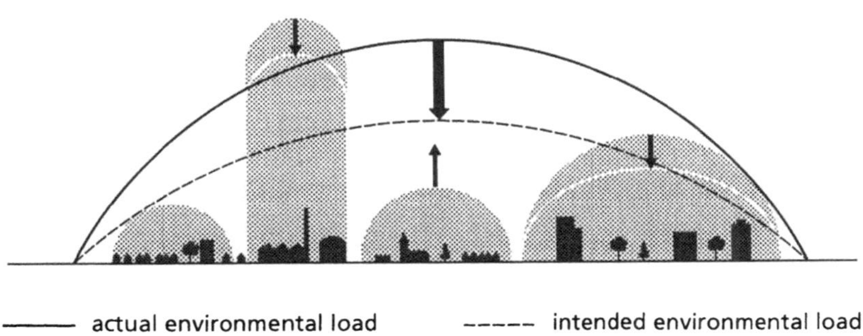

Figure 7.1: The 'dome' method (Humblet & De Roo, 1995, p. 38)

the city need to make an equally large contribution towards reducing the environmental load. Compensation between areas and environmental load categories is possible. For the time being, work is under way on the environmental aspects of noise, air pollution and external risks. The environmental load is indexed to a number between 0 and 100 and multiplied by the number of persons inconvenienced. This means that densely populated areas are more 'sensitive' than those in which fewer people live. Operationalisation of the dome concept was so complex that the method failed to be of any immediate practical value (De Roo, 1999).

7.8 The city and the environment method of approach

Partly on the basis of the observation that the environmental assessment methods discussed above offer too little solace for solving the bottlenecks between strict environmental standardisation on the one hand, and the need for new developments in cities on the other, another approach was developed in the 1990s which is referred to as 'the City and the Environment approach'. As is the case in policy innovations in the field of noise pollution and soil cleaning, the aspects of decentralisation and flexibility are central here too.

Leading up to the City and the Environment approach was the observation that the planning instruments of environmental policy and spatial planning were not fully in line with one another. The environmental standards for noise pollution, soil cleaning and air pollution – all experienced as stringent standards – often offer little opportunity for spatial developments such as housing and industrial premises because of the nuisance zones. The approach requires the step-by-step integration of environmental policy and spatial planning (Ministry of Housing, Spatial Planning and the Environment, 1995). The three steps involved are:

- *Step 1:* integrating environmental interests as early as possible in the spatial planning process.
- *Step 2:* by juggling with the scope available, using the scope that exists within the current regulations (environmental legislation as well as other relevant legislation).
- *Step 3:* deviating from the environmental standards under certain conditions: This means that in certain situations a higher level of environmental pollution is allowed than specified in the environmental standards.

The third step should be possible if it becomes evident that the first two steps have failed to achieve a good total result in terms of the 'optimal quality of life'. The aspect of 'compensation' plays an important role in this respect, the idea being that the 'environmental loss' in such a case must be compensated so that, on balance, the social climate in the area concerned is improved. The City and the Environment method of approach expresses a preference for compensation within that environmental compartment in which the 'loss' occurs. If this is out of the question, compensation may be sought in other compartments or even outside the direct environmental field, e.g. within the scope of green areas, the quality of housing and the surrounding areas, employment and public facilities.

Given that the general feeling was that it was too soon to embed the City and the Environment approach in legislation it was decided to start up 25 experiments to gain practical experience. These were projects focusing on the reconstruction of out-dated industrial estates, the revitalisation of post-war housing estates and the (re)construction of port areas and riverbank locations. The City and the Environment Experiment Act was introduced in 1998 to legitimate divergence from legislation in these experiments. The authorisation to diverge from legislation relates to legislation governing noise, air, soil and external safety.

Various requirements in terms of content and procedure are established in this Experiment Act regarding the actual bringing into practice of the authorisation to diverge from legislation. For instance, it is imperative that the first two steps have been taken before proceeding to take step 3. It must be clear that diverging from the relevant legislation is in the interests of the economic and efficient use of space, and that an optimal quality of life is achieved in comparison with the situation within which the prevailing regulations are used optimally within the scope of the current possibilities. The inspectorates for public health and spatial planning must also be consulted. Residents, the administrative bodies concerned and other stakeholders must be involved in the decision-making process (Staatsblad, 1998).

Most of the experiments were still engaged in steps 1 and 2 in 2000. The number of municipal authorities that will move on to step 3 will be limited. The initial experiences of the City and the Environment projects show that current legislation offers more scope than originally thought (Ministry of Housing, Spatial Planning and the Environment, 2000b). One of the main effects of City and the Environment is that it leads to a change in attitude within the municipality and thus to more collaboration, whereby the relevant parties refrain from limiting themselves to a single discipline. The step-by-step approach is generally found to be a useful method of approach by the municipal authorities involved given that it necessitates the consideration of environmental aspects in good time. Some municipal authorities indicated that they were already accustomed to working in this way before the experiments (Ministry of Housing, Spatial Planning and the Environment, 2000b).

Attention is given in these experiments to the involvement of societal actors in the policy-making process, albeit in widely different ways. However, the objective of this so-called 'open planning process', and the role of the various parties, is not always evident. Public servants and administrators often need to become accustomed to an open method of policy development. Other parties, such as residents and environmentalist groups, are also required to work in a different way because of the aspect of actively assisting in the thinking process (Ministry of Housing, Spatial Planning and the Environment, 2000b).

The approach taken in City and the Environment projects also has its limitations. Attention given to the environment is restricted to the aspect of the limiting conditions in the projects. Achieving a certain environmental quality is not generally incorporated in project development as a goal in itself. This has to do with the fact that the step-by-step approach was inspired by the limitations of environmental legislation. While the aspect of noise is given most attention in these experiments, the concepts of 'quality of life' and 'efficient use of space' have hardly been operationalised at all (Ministry of Housing, Spatial Planning and the Environment, 2000b).

The compensation option in City and the Environment projects has given rise to much debate. In practice, it is difficult for municipal authorities to operationalise the concept of compensation. The question asked is how far can you allow things to go, and whether, for instance, temporary compensation is an option (Ministry of Housing, Spatial Planning and the Environment, 2000b). Compensation within the same environmental compartment is difficult to realise in the case of soil pollution (Van Mourik, 1999). Nor is it always easy to make a difference between compensatory measures and measures that would be taken anyway.

More fundamental objections can also be mentioned. One general criticism on the idea of compensation is that it stimulates a calculating government that no longer focuses its energy on source-related measures; environmental issues are not solved in this way. Compensation can also give rise to tension in terms of social justice. By making compensation possible in other compartments or areas, a disproportionate relationship can arise between those persons that experience the non-achievement of environmental standards as a nuisance and the beneficiaries of that compensation. The question also arises as to which environmental values are assessed, and in what way. After all, environmental values are difficult to quantify and compare, one against the other.

7.9 Sustainable building

The City and the Environment method of approach discussed above is intended for the purpose of solving the tension between strict environmental standards on the one hand and the required spatial developments on the other by way of a well-considered process of project development in consultation with those concerned. There is, however, another theme that plays an important role in urban planning, that of sustainable building, which attempts to make a direct contribution towards a more sound urban development.

The concept of sustainable building has been established since the 1990s. This implies minimisation of the environmental impact (the use of energy, raw materials and water, the use of environmentally friendly materials) resulting from the construction and use of dwellings and other buildings. Sustainable building therefore relates to environmental problems in the city, as well as the environmental effects that occur outside the urban area. More recently, the concept of sustainable building has stressed the importance of spatial planning for the environment-friendly organisation of a district, for instance by creating traffic-abated areas, positioning dwellings in such a way as to facilitate the use of solar energy, and the establishment of marshland for purification purposes and ecological linking zones (Ministry of Housing, Spatial Planning and the Environment, 1999).

Elements of the traditional hierarchical management paradigm clearly play a role in policy on sustainable building. The Energy Performance Standard was introduced in 1995, setting out the minimum amount of energy efficiency project developers must meet in connection with housing and other building projects. While this standard has been a stimulus for the economic use of energy in dwellings, the examination of building plans against the prescribed Energy Performance Standard by municipal authorities still leaves a great deal to be desired (Van der Waals, 2001, p. 158). Subsidies also play a significant role in stimulating the implementation of environmental measures. Subsidies are made available because the cost of solar boilers, solar cells and water-saving measures is higher than the cost of conventional equipment.

In addition to traditional forms of management, policy places the emphasis on communicative instruments. Central government envisages more of a facilitating and stimulatory role set aside for itself than primarily a framework-setting role. A variety of instruments have been developed for the purpose of promoting integration of the environment and spatial planning into the field of sustainable building. For instance, checklists containing a large number of fixed and variable measures for materials, energy, water, raw materials and the indoor environment. These are the so-called 'national packages' for house building projects, property administrators, town planning, non-residential building projects and the

Civil and Hydraulic Engineering sector. These packages have no legal status but may be used by municipal authorities, town planners, project developers, housing corporations, architects and building contractors. The intention of these packages is to achieve unambiguity on necessary environmental measures for all parties involved in building projects.

Agreements have been drawn up on sustainable building at local and regional level between municipal authorities, project developers, housing corporations, architects, building contractors and, occasionally, energy suppliers and water companies. In 1998 a total of 350 such agreements were drawn up (Nationaal Dubo-Centrum, 1998), setting out the agreements reached on using 'national packages' or another list of environmental measures when carrying out building and renovation projects. A National Agreement on Sustainable Building was drawn up at national level between the government and umbrella organisations in the subsidised rented sector, in which an attempt is made to reach a 15% saving of energy over the period 1997-2001.

One specific communicative policy instrument is formed by the subsidised experiments and model projects in which more far-reaching environmental measures are applied. Nevertheless, experience with the communicative policy instruments is not always positive. The National Agreement on Sustainable Building and the regional agreements play little or no role in the decision-making process in building and renovation projects (Van der Waals, 2001, pp. 147-158).

Encouraging sustainable building is not always concerned primarily with the use of policy instruments but with organising collaboration processes between municipal authorities, project developers, housing corporations and energy suppliers (Van der Waals, 2001, pp. 147-158). Methods for shaping the participation of societal actors in new housing development projects and the restructuring of urban areas have, as yet, only been developed to a limited extent. One of these methods is the so-called 'CO_2 reduction workshop' focusing on one of the main aspects of sustainable building; the reduction of CO_2 emissions resulting from the consumption of energy and traffic. CO_2 reduction is often given too little attention in new housing development projects. In order to see that the reduction of CO_2 emissions is placed on the agenda at local level, TNO Strategy, Technology and Policy and Utrecht University developed a participatory method. This method consists of workshops which, as an experiment, were initially organised at two new housing development locations: the Eschmarke (in Enschede) and Bangert Oosterpolder (in Hoorn). Local civil servants, politicians, project developers, energy suppliers, transport companies and NGOs took part. The objective was to raise awareness for CO_2 reduction, to transfer and exchange knowledge, to involve relevant actors in advance, and to make arrangements regarding the ambitions and activities in the field of CO_2 reduction (Van der Waals, 2001).

The experiences with this form of participation were positive. The workshops evidently raise awareness and increase the knowledge of the actors with regard to the options for CO_2 reduction. Consensus was achieved on the ambitions for the level of CO_2 reduction and activities were organised to achieve that reduction at both new housing development locations where the policy experiments were conducted. Because of the length of time involved in new housing development projects, the effects on the eventual implementation of the options cannot be determined as yet. And yet it is quite clear that far less has been done with the results in Eschmarke than in Bangert Oosterpolder, possible reasons being: the project manager's role, the more closed style of policy making in Eschmarke, plus the fact that the participants in Bangert Oosterpolder regarded CO_2 reduction as an important subject before the workshop. Workshops are apparently a useful form of process management if linked with a specific implementation trajectory (Van Hoorn et al., 2001).

Experience with the different forms of management in sustainable building policy shows that success can only be achieved through a combination of management at central level and a process approach being taken at the local level. Process management at local level is necessary to motivate the actors to use those options that cannot yet be made obligatory, and to be able to use the most up-to-date knowledge and technologies. However, a decentralised approach based on the voluntariness and initiatives of local actors has its limitations too. For instance, it appears that sustainable building is given too little attention when restructuring urban areas. The implementation of sustainable building measures apparently depends to a large extent on (the quality of) the process management of municipal authorities or other actors (Van der Waals, 2001, pp. 147-158). It is therefore appropriate that the government supports the optimum use of knowledge and experience. Central management by means of regulations and subsidies is also essential to make new options financially feasible, and – through legislation – to enforce a minimum level of energy economy among those actors who, if the approach is voluntary, are difficult to reach and influence (Van der Waals, 2001, pp. 195-206).

7.10 Local Agenda 21 in the Netherlands

The last subject we wish to discuss is not so much a separate field of policy but a policy initiative that plays a significant role in the urban environmental policies of many nations: Local Agenda 21. Local Agenda 21 stemmed from the 1992 United Nations Conference on the Environment and Development (UNCED) held in Rio de Janeiro. According to Agenda 21 (which was signed at the conference) local authorities must, in consultation with residents, businesses and NGOs, develop a Local Agenda 21 focusing on a sustainable society. Local Agenda 21 is therefore in line with the trend towards a higher level of participation of the societal actors in environmental policy.

More than 25% of all Dutch municipal authorities – about 150 – had a Local Agenda 21 in place in 1998. One of the most significant motives for introducing Local Agenda 21 was the subsidies that were available for setting up such an agenda by the Ministry of Housing, Spatial Planning and the Environment. Local Agenda 21 also had the support of the Committee on International Co-operation and Sustainable Development, a platform of NGOs involved in drawing up the annual Dutch reports on Agenda 21 (Andringa, 1998).

Local Agenda 21 has been less of a success in the Netherlands than in many other countries. There are many reasons why: a large amount of environmental policy has already been developed in the Netherlands at the local level and thus the added value of Local Agenda 21 has less of an impact. Furthermore, the 'VNG' *(Association of Netherlands Municipalities)* has done little to encourage Local Agenda 21. The VNG is not as active in this respect as similar organisations in other countries (Bouvy, 1996).

There are no firm rules for organising a dialogue on Local Agenda 21. Consequently, several different procedures have ensued. In many cases, activities are developed which offer citizens the opportunity to contribute subjects and ideas for concrete action, but with no clear insight as to what the municipality will do with the input. This is often accompanied by the use of incidental forms of dialogue, such as a workshop or a debate where everyone can have their say. Some municipal authorities also conduct surveys or interviews among their residents. However, a more prolonged long-term dialogue in which the parties can confront one another with their different points of view and try to reach solutions by improving their understanding of one another's views is usually lacking (Vermeulen et al., 1999).

One of the most talked-about examples in the Netherlands is the approach followed in The Hague. Work began on this in September 1995, and the Local Agenda 21 was presented in December 1996. In October 1996 The Hague municipal authorities were awarded a European Environment Prize for their Local Agenda 21.

A bottom-up approach was adopted in The Hague, which took the wishes and desires of the public as the point of departure. This implied that the municipality of The Hague was merely one of the parties involved. The municipal dialogue consisted of the following elements: informing the city (the "shop window"), the discussion ("the city debate") and the implementation ("doing it"). The organisation consisted of a steering group, a project organisation, theme groups and working parties. The steering group, consisting of members from various segments of society, provided a link between the city council and the public and had to supervise the debate. The task of the project group was to stimulate and support the theme groups, which for their part were required to come up with creative solutions. There were eight theme groups for global relations, energy, waste and raw materials, traffic and transport, nature and landscape, sustainable building and living, communication and neighbourhood initiatives. In addition to members of the public, these theme groups also consisted of civil servants who took part in a personal capacity. The working parties, to which anyone could join, were responsible for implementing the ideas. The purpose of the working parties was that they should work out ideas or projects which could contribute to a sustainable The Hague. The members of the public were kept informed via the Internet, newsletters and public meetings (Vermeulen et al., 1999; Gemeente Den Haag, 1996).

Some 3,000 people and 50 organisations were involved to some extent in the The Hague Local Agenda 21. In total, some 70 'ideas' were generated, some of which have meanwhile been implemented, such as the Green Pages, the construction of a bicycle shelter, the sale of thousands of eco flower bulbs, a conference on 'Mobiliteit op Maat' *(Customised Mobility)* and the 'green against graffiti' idea (covering walls with climbing plants to combat graffiti).

And yet there are also several critical comments that can be made regarding the approach taken by the The Hague authorities. The influence Local Agenda 21 had on the municipal organisation as a whole has, for the time being, remained limited. This is connected with the fact that Local Agenda 21 is seen as a project which must be completed within a specific period of time. Moreover, most of the participants in the debates have already become involved in the policy-making process. No real 'laymen' were reached. Nor was there any success in involving trade and industry with Local Agenda 21 (Andringa, 1998).

Similar problems also face other municipal authorities. In general, the link with existing decision-making procedures is limited and thus drawing up a Local Agenda 21 becomes an isolated activity. Trade and industry is not often involved in Local Agenda 21. Moreover, municipal authorities mainly use Local Agenda 21 to discuss the actual liveability of the immediate surroundings; the global aspects are given little attention. The limited effect on regular policy processes is also a problem for many other uses of the participatory implementation of policy on urban environmental issues (Vermeulen et al., 1999).

Across the board, Local Agenda 21 has failed to result in a permanent dialogue between the city's inhabitants and the municipal authorities. Consultative platforms established for Local Agenda 21 often lead a languorous existence or are a meeting place for the 'inner circle' of environmentalist groups and municipal officers. However, the question is how problematic this actually is. There are many other forms of consultation, participation and collaboration, which basically correspond with the points of departure of Local Agenda 21 (Smits, 2000).

The feeling that the integration of Local Agenda 21 does not run all that smoothly in municipal policy is also experienced in other countries. In the UK we see that politicians and civil servants have difficulty in integrating the aspect of sustainability in day-to-day policy, and that Local Agenda 21 is seen as a separate policy process (Doak, 1998). According to Lustig & Weiland (1998) the municipal authorities in Germany focus more on implementing regulations than on incorporating ideas from outside the municipal organisation into policy. It is also evidently difficult to involve certain groups such as women, the youth, the elderly and the poor in Local Agenda 21. Budgetary obstructions are also apparent in the municipalities (Lustig & Weiland, 1998).

7.11 Conclusions

In this chapter we have illustrated that the paradigm shift in environmental policy on urban areas is characterised by a decentralisation of the decision-making process, the integration of environmental policy and spatial planning, and the participation of citizens, organisations and businesses in policy. To a large extent these policy innovations are the outcome of the limitations of hierarchical management strategy as has been pursued for many years in the policies on noise abatement and soil cleaning. It is apparent that this approach comes up against problems when implementing policy: the societal consequences of the strict implementation of environmental standards are, at least in several cases, deemed unacceptable. Also, wider trends in government policy, especially the decentralisation of authority and simplification of legislation, are a stimulus for policy innovation. While each of the three policy developments offer advantages and new opportunities, they also have their drawbacks and associated risks.

In order to promote the practicability of and the support for environmental policy, the aspect of decentralisation offers the opportunity to make concessions in certain situations regarding sectoral environmental goals. Account can be taken with local circumstances, such as the (presumed) inhabitants' sensitivity to environmental problems and the desired spatial development. However, the opposite can be the 'snowing under' of environmental goals in the decision-making process adopted by municipal authorities. There is a much stronger alliance with the local economic interests in these authorities than among the higher authorities.

Furthermore, decentralisation does not imply that national legislation is no longer valid. Environmental standards are however made more flexible, and the emphasis is placed on procedural requirements. Additionally, national and European environmental policy remains to be of essential importance for separate aspects of policy such as policy on mobility, building regulations, energy-saving and environmental management in industry.

The intention of decentralisation is partly to facilitate the integration of environmental policy and spatial planning, the second policy innovation. Partial integration is already the case given that the two fields of policy, each with their own legal framework and system of plans, continue to co-exist. Nevertheless, environmental standards are weighed against other societal interests; they are not automatically seen as the limiting conditions for spatial planning. This makes the question 'how disproportionate aspects of quality should be weighed' more pivotal. The concept of compensation has become more succinct as a concept for shaping this weighing up procedure, and yet it is still apparently difficult to operationalise. One objection against this concept is that it can be used to disguise the fact that environmental problems are basically not being solved.

The third trend, that of the participation of citizens, businesses and NGOs in policy, is partly linked to the first two policy innovations and becomes manifest in several different ways. The participation of citizens and organisations is felt to be important in policy on noise abatement and in City and the Environment projects because these groups are recognised as stakeholders; their involvement has even been embedded in law. The participation of market parties in the financing of soil clean-up projects is pursued in policy on soil abatement. Participation of citizens and organisations in Local Agenda 21 is more often focused on generating ideas. The trend towards more participation can be appreciated much more from the viewpoint of supporting policy, making use of the creativity element, and giving consideration to the relevant interests. However, the organisation of open planning processes and abandoning the focus on internal decision-making processes is not an easy thing for all municipal authorities to achieve. Especially the linkage of pursuing different sorts of participatory policy to existing formal policy frameworks must be given additional thought.

The question as to how far the policy innovations will lead to a reduction in the environmental load is difficult to answer, particularly because it is far too early to drawn conclusions in this respect. Only time will show how the local authorities have filled in the details of their own scope for policy making within the revised policy frameworks. However, at this point it can be observed that certain aspects of urban environmental problems have hardly been addressed in the policy. For instance, the 'distribution' of environmental problems throughout the community has still not been investigated in environmental policy. Additionally, little has been done about bringing the accumulated urban environmental load into the picture. Moreover, the experiences with Local Agenda 21 show that local authorities focus more on the liveability of the immediate surroundings and less on the global aspects.

Finally, the significance of the paradigm shift in environmental policy can be established by investigating how it relates to the societal causes of urban environmental problems such as the rise in level of mobility, economic growth and smaller households. The former paradigm apparently allowed less scope for such developments, but the implementation failed. In the new paradigm, more is being done to meet market trends, especially to promote the implementation of policy. Nevertheless, in a certain sense we still see a strong continuation of the same policy. The societal developments mentioned in the foregoing, both in the former and in the new paradigm, have not, or to only a small extent, been discouraged as such, policy being aimed at limiting their detrimental effects. Therefore, the importance of a change in the organisation of policy with regard to how the environmental load will develop must, to a certain extent, be brought into perspective.

References

Anderson, W.P., P.S. Kanaroglou & E.J. Miller (1996) 'Urban Form, Energy and the Environment: A Review of Issues, Evidence and Policy', *Urban Studies,* Vol. 33, No. 1, pp. 7-35.

Andringa, J. (1998) 'The Influence of Local Agenda 21 on Local Policy and the Quality of Decision-Making: The Pioneer City of The Hague', in: F.H.J.M. Coenen, D. Huitema & L.J. O'Toole. (eds.), *Participation and the Quality of Environmental Decision Making,* Kluwer Academic Publishers, Dordrecht, pp. 107-123.

Borst, H., G. de Roo, H. Voogd & H. van der Werf (1995) *Milieuzones in beweging. Eisen, wensen, consequenties en alternatieven,* Samsom H.D. Tjeenk Willink, Alphen aan den Rijn.

Bouvy, F. (1996) *Duurzame ontwikkeling: een collectief belang. Een onderzoek naar de participatie van maatschappelijke organisaties in een Lokale Agenda 21-proces, Afstudeerscriptie Milieukunde,* Universiteit Utrecht, Utrecht.

Consultancy Service for Traffic and Transport (AVV) (1997) *Infrastructurele ontwikkelingen 1997,* Rotterdam, pp. 9-28.

De Roo, G. (1999) *Planning per se, planning per saldo. Over conflicten, complexiteit en besluitvorming in de milieuplanning,* Sdu Uitgeverij, Den Haag.

Doak, J. (1998) 'Changing the World Through Participative Action: The Dynamics and Potential of Local Agenda 21'. in: F.H.J.M. Coenen, D. Huitema & L.J. O'Toole (eds.) *Participation and the Quality of Environmental Decision Making,* Kluwer Academic Publishers, Dordrecht, pp. 77-91.

Farjon, J.M.J., N.F.C. Hazendonk & W.J.C. Hoeffnagel (red.) (1997) *Verkenning natuur en verstedelijking 1995-2020,* Achtergronddocument 10 Natuurverkenning '97, IKC Natuurbeheer, Wageningen.

Gemeente Den Haag (1996) *Onze toekomst in eigen hand,* Den Haag.

Humblet, A.G.M. & G. de Roo (red.) (1995) *Afstemming door inzicht, een analyse van gebiedsgerichte milieubeoordelingsmethoden ten behoeve van planologische keuzes,* Geo Pers, Groningen.

Hut, J.A. & G. de Roo (1999) 'Decentralisatie van bodemsaneringstaken', *ROM Magazine,* jaargang 17, nr. 9, pp. 2-5.

Interdepartemental Study Group on Soil Cleaning (Interdepartementale werkgroep bodemsanering) (1997) *Gerede grond voor groei. Nieuwe impulsen voor de bodemsanering,* Den Haag.

Lustig, S.H. & U. Weiland (1998) 'Learning from past experience? Local Agenda 21 processes and integrated urban development planning in Germany', in: F.H.J.M. Coenen, D. Huitema & L.J. O'Toole (eds.) *Participation and the Quality of Environmental Decision Making,* Kluwer Academic Publishers, Dordrecht, pp. 93-106.

Ministry of Housing, Spatial Planning and the Environment (Ministerie van VROM) (1995) *Stad & Milieu Rapportage. Waar velen willen zijn, is ook een weg,* Den Haag.

Ministry of Housing, Spatial Planning and the Environment (Ministerie van VROM) (1999) *Beleidsprogramma Duurzaam Bouwen 2000-2004. Duurzaam verankerd,* Den Haag.

Ministry of Housing, Spatial Planning and the Environment (Ministerie van VROM) (2000a) *Vijfde nota over de ruimtelijke ordening 2000-2020; Ruimte maken, ruimte delen,* Den Haag.

Ministry of Housing, Spatial Planning and the Environment (Ministerie van VROM) (2000b) *Verdiepingsslag Stad & Milieu,* Den Haag.

Nationaal Dubo-Centrum (1998) *Dubo-convenanten in ontwikkeling; een handleiding voor en informatie uit de praktijk,* Aeneas, Best.

National Institute of Public Health and the Environment (RIVM) (1997) *Nationale Milieuverkenning 4 1997-2020.* Samsom H.D. Tjeenk Willink, Alphen aan den Rijn.

National Institute of Public Health and the Environment (RIVM) (1998) *Milieubalans 98. Het Nederlandse milieu verklaard,* Samsom H.D. Tjeenk Willink, Alphen aan den Rijn.

National Institute of Public Health and the Environment (RIVM) (2000) *Milieubalans 2000. Het Nederlandse milieu verklaard,* Samsom H.D. Tjeenk Willink, Alphen aan den Rijn.
Regiegroep Bever (2001) *Eindrapport BEVER/UPR: de veranderingen in het bodemsaneringsbeleid als uitwerking van het Kabinetsstandpunt van 1997,* Den Haag.
Smits, J.M. (2000) 'Gemeenten en beleidsuitvoering', in: Driessen, P.P.J. & P. Glasbergen (red.) *Milieu, samenleving en beleid,* Elsevier bedrijfsinformatie, Den Haag, pp. 187-209.
Staatsblad (1998) *Wet van 26 november 1998, houdende regels over experimenten inzake zuinig en doelmatig ruimtegebruik en optimale leefkwaliteit in stedelijk gebied (Experimentenwet Stad en Milieu),* no. 684.
Van der Vegt, C. & W.J.J. Manshanden (1996) *Steden en stadsgewesten, economische ontwikkelingen 1970-2015,* Stichting voor Economisch Onderzoek der Universiteit van Amsterdam. Sdu Uitgeverij, Den Haag.
Van der Waals, J.F.M. (2000) 'The Compact City and the Environment: A Review', *TESG,* Vol. 91, No. 2, pp. 111-121.
Van der Waals, J.F.M. (2001) *CO_2-Reduction in Housing, Experiences in Building and Urban Renewal Projects in the Netherlands,* Rozenberg Publishers, Amsterdam.
Van Hoorn, T.M.M., J.F.M. van der Waals, W.J.V. Vermeulen & A.F.L. Slob (2001) *CO_2 Reduction in New Housing Estates: A Workshop as a Building Bloc,* National Research Programme on Global Air Pollution and Climate Change, Bilthoven.
Van Mourik, G.L. (1999) *Stad en Milieu. Een vergelijkend onderzoek van enkele experimenten,* Afstudeerscriptie Milieukunde, Open Universiteit, Heerlen.
Verdaas, J.C. & P. Glasbergen (1997) 'Stedelijk beheer en de duurzame stad', in: *Handboek Stedelijk Beheer,* VUGA Uitgeverij, Den Haag, pp. 71-105.
Vermeulen, W.J.V., J.F.M. van der Waals & P. Glasbergen (1999) 'Public Participation in Planning of Urban Sustainable Development: An Analysis of Recent Practice', in: E. Wever (ed.), *Cities in Perspective,* Van Gorcum, Assen, pp. 159-177.
Verroen. E.J., H.D. Hilbers & C.A. Smits (1995) *Modeltoets Randstadvisie: de resultaten,* TNO-INRO, rapport 95/NV/124, Delft.
VNG/IPO/Ministerie van VROM (1998) *Beleidsnota vernieuwing geluidhinderbeleid, een nieuwe sturingsfilosofie voor het toekomstige geluidbeleid,* Den Haag.

PART 4

POLITICAL INSTITUTIONAL ASPECTS

Chapter 8

ENVIRONMENT AND PARTICIPATION

The shifting significance of a double concept

Pieter Leroy and Jan P.M. van Tatenhove

8.1 Introduction

Since the rise in environmental awareness at the end of the 1960s and early 1970s it would appear that the environment and participation are two inextricably linked aspects of the same 'green discontent'. Since about 1970 that discontent has gradually increased, supported by a growing environmental concern, and reinforced by successive environmental conflicts. There were always two aspects that went hand in hand with that green discontent: the protests always concerned the environmental load of planned decisions ánd the way in which those decisions were taken. This double discontent was expressed when polluting industries or individual businesses were established, when infrastructural work was undertaken, with the construction of new roads etc. Citizens and the environmentalist movement voiced their objections to the actual content of the decisions concerned because of the environmental impact involved, as well as the way in which the decisions had been taken, particularly because of the lack of public participation. Exactly the same double discontent is being expressed in the environmental protests made today, demonstrations against airport expansions everywhere in Europe, for instance, or the protests voiced against 'globalisation': objections against the actual content of decisions go hand in hand with objections against the non-participatory manner in which they are reached.

P.P.J. Driessen and P. Glasbergen (eds.), Greening Society, 163–184.

8.2 The political message behind 'the green discontent'

Environmental issues were not the only ones in which participation was called for in the late 1960s and early 1970s. The green discontent in those years was part of a wider public criticism; criticism focusing on the capitalist system and the role of the state which maintained inequality. The call for the politicisation of capitalist institutions, for the democratisation of the state, and for the emancipation of oppressed groups was interpreted in radical arguments for more participation in virtually all sections of society: in churches and in trade unions, at universities and in politics. Calls for a change in the social structure and for the more active involvement of citizens, members, co-workers, students, etc., in decision-making processes, were heard everywhere. The early 1970s can therefore be typified as a period of political radicalisation. This radicalisation was symptomatic for the then demonstrably diminishing legitimacy of a variety of traditional institutions, such as the churches, the universities, the trade union movement, politics, etc. Counter-movements came into being in each of these sectors, challenging the legitimacy of the existing institutions, arguing for other, particularly more participatory structures. Each in their own particular way, the environmentalist movement, the women's movement, the students' movement and numerous other 'new social movements' propagated a comparable, often radical body of ideas in which the old institutions were expected to be eliminated and replaced by more modern organisations.

Whereas the demand for a higher level of participation among citizens, students, women and other groups resounded in many sectors, we restrict ourselves here to the (demand for) political participation only. This refers to the participation of citizens and social groups in political decision-making processes and the realisation of government policy. Defined in this way political participation is nothing new, nor was it new in the 1970s: after all, people in democratic states have for some time now been accustomed to the right to vote, the right of petition, the right to referendum and the right as an interested party, as a citizen, to object to certain government plans. These forms of political participation are familiar. The same applies to the participation of non-governmental organisations: not only political parties but also trade unions, employers' organisations, organisations in the field of welfare work, culture, etc., are all directly or indirectly involved in the preparation or implementation of government policy. The government frequently invites these groups to join advisory and consultative bodies, and from there they help to formulate new policy. Even when the degree of involvement of these non-governmental groups and organisations differs from one country to another, varying from simply giving advice to helping implement policy, 'participation' is at stake in one way or another everywhere.

However, the call for political participation in the late 1960s and early 1970s was for new, more radical forms of participation, for a more direct influence on policy. The representative democracy was (or had become) too indirect and partly overruled by forms of (neo)corporatism in which a small number of interest groups in actual fact ran the whole show, important decisions being made after consultation among the elite of these groups (see Chapter 6). It was that tradition and structure of elitist consultation as the basis of the decision-making process in particular that silenced the citizens: the constant production of political compromises between the elite of the main political and socio-economic groups could only be maintained by a simultaneous and far-reaching depoliticisation of politics on the one hand, and by keeping one's own followers at a distance and demobilising them on the other. This form of politics and decision making had to either be replaced or at least supplemented with more contemporary forms of participation. The demands for more political participation also varied in their degree of radicalness: ranging from moderate requests for additions to or innovation of the representative democracy to fierce arguments for a basic, democratic policy. There was also a difference of opinion seen among the members of the environmentalist movement, as we shall see below. Nevertheless, participation was still the keyword.

However great the difference in level of radicalness, the call for participation implied a demand for innovation in terms of both content and organisation. This applied in particular with regard to how environmental issues were tackled. Especially because of the then emerging awareness of the environment and environmental policy, and because environmental issues relate to the quality of everybody's immediate environment, the development of this field of policy was closely interwoven with the call for participation. As already said, this also applies with regard to the environmental protests made after the 1970s: concern for the environment and the demand for more participation was also expressed in the green discontent in this respect too. The rest of this chapter deals with the questions how the environment and participation have apparently been inextricably linked over thirty years of environmental debate, and why in those thirty years the attitudes and practises regarding participation have changed quite substantially.

8.3 Participation enforced and gradually institutionalised

Around 1970 many western nations faced a series of environmental conflicts. Virtually every European country has its own list of national, and occasionally international, prominent environmental battle fields dating from the 1970s and 1980s, as also do the USA, Canada and Japan. These were sometimes intense forms of air or water pollution, especially of industrial origin (London, Liège, Lake Tahoe), occasionally environmental incidents (the Torrey Canyon in Brittany, Seveso, Bhopal), and sometimes the establishment of new businesses (Progil in Amsterdam, and virtually all (potential) nuclear plant locations), the expansion or construction of new infrastructure (Frankfurt, Tokyo). Many of these cases have been studied and described in an often moving way, for instance in Castells (1974), Gladwin (1980), Blowers (1984) and more recently, Gould et al., (1996). However different these cases might be and how diverse the perspectives that were at the base of the studies, there is still one striking, common element, and that is the two aspects of green discontent in each of them: protest among citizens, local residents, the environmentalist movement, and other opponents always relates to the underestimation of the environmental effects of the action in dispute, and to the insufficiently democratic character of the decision-making process regarding that action.

The first complaint – too little attention given to the environmental impact – led to the gradual development and institutionalisation of environmental policy (see also other chapters of this book). It also led to the tightening up of existing instruments and the development of new ones to safeguard the interests of the environment. We will give two international examples to illustrate this. First, a 'Nuisance Act' was in force virtually everywhere in Europe in the early 1970s. Whereas the specifics of these acts differed according to the national institutional structures and cultural traditions, the principle was more or less the same: a permit was required for a large number of hazardous or objectionable activities, usually businesses. These permits originally related to danger, damage and nuisance outside the business, and later on, inside (i.e. on work safety and working conditions). Since the 1970s, especially under the influence of environmental protests, comparable changes have been observed in this legislation in all countries. In the first place, the scope of the legislation has become constantly wider or legislation was broken down to cover the different pollution categories. Secondly, the scope of the term 'nuisance' was also subjected to a constant widening: instead of only potential danger, damage and nuisance for the immediate surroundings, i.e. human beings in particular, the potential damage to the environment was also gradually taken into consideration. The use of raw materials and energy is now taken into consideration in some countries when assessing permit applications. And thirdly, openness in the application procedure and the decisions involved in granting such permits are gradually being intensified to a differing extent from one country to another. We shall return to this later.

A second example of innovations in the realm of environmental instruments is the environmental impact assessment or EIA. Since the early 1970s a great deal of environmental protest has been aimed at the fact that the environmental effects of a planned action were only taken into consideration at too late a stage in the decision-making process. Or worse still, that they had hardly been looked at at all and consequently their full impact had only become apparent at a later date. Hence the call to study the environmental effects of planned actions prior to the decision-making process in an environmental impact assessment and to take the results of such an assessment into consideration in the decision-making process. Environmental impact assessments, developed in the USA and Canada at first and subsequently tested in the Scandinavian countries, the Netherlands and Germany, have been an institutionalised component of environmental policy in all EU Member States and many other countries since 1985. It was in fact one of the first preventive environmental policy instruments; moreover, it was one of the first instruments in which a broad, integral meaning was given to the concept of 'the environment'.

In short, the first demand expressed in environmental protests, to keep a better eye on the environmental effects of various activities, led to such an innovation of environmental policy instruments that 'the environmental interest' was operationalised and taken care of better. The way in which this institutionalisation took place, the changes in and the effectiveness of that environmental policy are dealt with in detail elsewhere in this book.

The second demand ensuing from the green discontent, the demand for more participation, also resulted in adjustments being made. As explained earlier, there was a gradual increase in the opportunities for participation in the existing regulations under public law among local residents and other stakeholders with regard to environmental permit systems for instance. This, by the way, applied not only to the emerging environmental policy, but also to the existing regulations on spatial planning. At the same time we see new, more participatory forms of policy emerging, certainly in the field of spatial planning (falling under the scope of terms such as structural planning and regional planning). Also in environmental policy there was an easing of the procedures for lodging objections and appeals, and information was provided at public hearings, panel discussions etc. Environmental impact assessment is given here again as a typical example. Virtually across the whole of Europe have environmental impact assessments resulted in more opportunities for public participation and

EIA-procedures are now equipped with various forms of information and participation.

8.4 Nuclear energy: test case and an obstacle in the way of more participation

This widening of the possibilities available for political participation does not yet mean the elimination of 'green discontent'. As indicated above, some groups argue for more radical action in public administration. After all, expanding participation in the permit-granting process for a certain activity or industry does not automatically imply a higher level of participation. In the first place it remained to be seen exactly what the actual influence of that participation would be, and secondly, it was still only a matter of decisions taken at the level of individual businesses or their branches, and not the environmental risks involved for an entire branch of industry. The nuclear energy issue reflected this problem everywhere in Europe.

In the years between 1973 and 1986, the most controversial topic in virtually every western country regarding the environment was nuclear energy. The subject of energy had become topical since the oil crisis in 1973-1974. Considerable price increases were also seen in 1979-1981. Nuclear energy was put forward as a political and technically safe alternative for the high level of dependency on oil from the Middle East (at that time there was no question of oil being extracted from the North Sea). And yet nuclear energy gave rise to enormous opposition in society. Those opposing nuclear energy not only associated it with the atomic bomb and the arms race, but also saw it as symbolic of a large-scale technology dominated by technocrats, which the man in the street had absolutely no grip on. In short, nuclear energy was dangerous, unsafe, undemocratic and consequently politically and socially unacceptable.

Nuclear energy in general, and the planned construction of nuclear power stations or other nuclear plants in particular, led to fierce political controversies in the whole of Western Europe and to physical battles being fought out between the environmentalist movement and the police. Well-known are the confrontations as a result of Creys-Malville (France), Brokdorf, Wyhl, Gundremmingen and Kalkar (Germany). Later, it was the planned locations for nuclear waste that became the arena of turbulent environmental conflicts, such as Elstow (UK) and Wackersdorf and Gorleben (Germany) (Kitschelt, 1986; Blowers, Lowry & Solomon, 1991).

Now, fifteen to twenty years later, we are able to draw up the balance of the effectiveness of all those activities and protests. Some nations actually dropped the idea of nuclear energy immediately, often because they had other adequate sources of energy, others simply brushed the demonstrations aside and, particularly on military and technological grounds, gambled heavily on the nuclear energy card. In both cases it would seem that the influence of the demonstrations was relatively limited. This was partly because while the politico-institutional structure offered scope for 'participation' in the individual dossiers of business establishments (with only a limited impact, see below), there were no participatory instruments available for a political debate on the principle of nuclear energy itself. Conversely, the decision-making process on nuclear energy was for many people – and still is – a symbol of an autocratic style of decision making which is dominated by political, military and technological considerations. It must be ascertained here that the environmentalist movement failed to succeed in breaking through that decision-making process or in making it more participatory.

Nevertheless, nuclear energy was reason for several European nations to experiment with more participatory decision-making scenarios at national level. The controversial character of nuclear energy and the fierce demonstrations against it, including those at Almelo (1978), Kalkar and particularly Dodewaard (1980-1981), forced the Dutch Government to adopt a separate decision-making procedure. When all was said and done, the government realised that it was confronted with fierce environmental protest and a rapidly diminishing faith in the public opinion on nuclear energy, which was changing into distrust and opposition. Nuclear energy, particularly in a country with a tradition of large-scale political legitimacy and consensus, was threatening to lead to a legitimacy crisis. This threat was also experienced as such in politics, even in the conservative segment. It was partly because of this that the then centre-right government, the Van Agt government, brought a halt to the procedure for a key national (physical) planning decision on nuclear energy in 1978 which was accompanied by an elaborate, but somewhat one-sided publicity campaign. Instead, it chose for a wide-ranging public debate: meetings were held across the country in order to give the public the opportunity to participate, public hearings and discussions were organised, preceded by extensive information and communication on nuclear energy, leading to a widely supported 'decision'. This public debate was an entirely innovative element in the organisation of political participation at national level: a form of direct democracy was introduced in addition to, or in this particular case – preceding – the decision-making process through the representative democratic bodies. We shall discuss the relation between the two later.

The actual impact of this wide-ranging public debate on nuclear energy, however, was insignificant. After all, a large majority of Dutch citizens was apparently against the further expansion of nuclear energy. This did not however stop the next centre-right government from still making a choice for nuclear energy some years later (1985) in the key national (physical) planning decision taken at that time. The main arguments for this were (a) a different estimation of the economic benefits and environmental risks, (b) the fear that the Netherlands would miss out on the developments in this technology, and (c) the contribution nuclear energy would make to combating the 'acid rain' that had become the focus of interest in the early-eighties. This government decision led to an enormous amount of anger and disillusionment among the general public and the environmentalist movement in particular. After all, the government had taken a decision that was against the very clear wishes of the majority of the population. The then minister for the environment, Winsemius, said on this matter: "It is not that we have disregarded the outcome of the public debate (...). But we have reached a different conclusion on the basis of sound arguments. No single participant in the public debate needs to feel ignored. We have never given the impression that the outcome of the public debate would be binding" (Nauta & Steenbrink, 1985). This quotation clearly reflects the unresolved tension between the institutionalised, constitutional representative democracy on the one hand, and the more or less experimental addition thereto. Shortly after the Dutch Government had taken the decision to go ahead with nuclear energy, the dramatic accident in the nuclear power plant at Chernobyl occurred (April 1986). This led initially to the postponement of the decision and finally to the suspension of its implementation. The gradual discontinuation of nuclear energy in the Netherlands was chosen for after 1996 when the decision was taken to close down the Dodewaard and Borssele plants. As is known, the coming into power of the present red-green government in Germany has also now resulted in the decision (2000-2001) to gradually bring a halt to nuclear energy there too. This can be seen as one of the long-term effects of the environmental protests of the 1970s and 1980s. At the same time we must also ascertain that this German decision was also made via the constitutional political circuit, and not via the participatory innovations that were called for at the time.

8.5 The effects of the new forms of participation

Since as far back as the 1970s, study has been conducted into the (planned and unintentional) effects of these new instruments for more participation in environmental policy, spatial policy and other associated areas of policy. First of all, these studies have confirmed that these new instruments, despite all the good intentions, failed to lead to more or better political participation. The absent citizen – the actual reason for the demands for participation – remained absent. Secondly, the new set of participation instruments failed to countervail the mechanism of cumulative participation: individuals and groups that already had a political involvement – by virtue of their interests, status, knowledge, familiarity etc., – were given extra opportunities to participate, while the less interested citizen was not sensibilised by the new opportunities, let alone mobilised. While public hearings and panel discussions were attended by environmentalist movement professionals, the 'man in the street', the citizen they professed to represent, was still absent. This, thirdly, had to do with the limited amount of actual influence of the new participation opportunities on the decision-making process. The dominant impression among citizens and the environmentalist movement was that the government tended to frequently ignore the outcomes of public inquiry procedures and participation procedures. Environmental objections were simply discarded as being localistic, inspired on (local) internal interests (or: on the so-called NIMBY reflex), and were inadequately underpinned scientifically. Moreover, the power base of the established ministries and the major interest groups had apparently been left unaffected by the opportunities to comment. This was evident from numerous local environmental dossiers (Huberts, 1988), it was, as has already been pointed out, mainly evident in dossiers of national or international importance, such as nuclear energy, the intensification of air traffic and other large-scale and internationally relevant infrastructure (Lowe & Goyder, 1983; Kitschelt, 1986). In short, the efforts of the citizens and the environmentalist movement to make maximum use of the participation opportunities that were offered failed to lead to a change in the balance of power, while this was one of the – radical or reformistic – 'ambitions' of the societal discontent. Of particular note here was that despite all demands and arguments for a more direct form of participation, the representative democracy still had the final word. In other words, the new instruments for more political participation had brought about no change in respect of the existing political balances of power, nor with regard to government and parliament's responsibility. In other words: while political participation was, and still is, 'accepted' by the representative democracy as an addition, it is not seen as a rectification, let alone as a substitute.

Other conclusions of the study into the effects of the new forms of participation were related to the unintentional effects thereof, especially the impact on the role and strategy pursued by the environmentalist movement. After all, while the citizens remained absent, the organised environmentalist movement in particular (on behalf of the citizen?) made use of the new set of participation instruments. Enforcing these instruments was to a large extent to the merit of the environmentalist movement. Yet in turn, the institutionalisation of those participation instruments had an – unintentional – influence on its position and strategy.

This calls for an explanation. The rise and development, the political diversity and the strategies and influence of the environmentalist movement (in different nations) have been described quite adequately elsewhere (Lowe & Goyder, 1983; Jamison et al. 1991; Kriesi et al., 1992). Here, we discuss only the position and role of the environmentalist movement as the advocate of and vehicle for participation. As said, enforcing more political participation in the field of the environment was to a large extent to the merit of the environmentalist movement. It was also by far the most important user of these instruments and was therefore a link and vehicle of participation between the citizen and the government. Nevertheless, there was not exactly unanimity regarding that role: as was the case in other new social movements in the 1970s, there was discord in the environmentalist movement too about the amount of radicalness to be used for demanding political participation, and about the additional, corrective or even substituent character of that participation as opposed to the existing representative democracy. However, as the political opportunity approach illustrated, the strategic preferences and options of the environmentalist movement are connected with the political opportunities offered by a country's institutional structure. The absence of political opportunities in what was then Western Germany played a significant role in the radicalisation of the German environmentalist movement as a component of the so-called *'ausserparlementarische Opposition'* and in the strategic choice to establish a political formation. Conversely, the sharp dichotomy of the British and the French political landscape – the one with a two-party system and the other with a traditional left-right contrast – are important explanatory factors for the lack of success of the green parties in these two countries. Also the rapid acknowledgement of the environmentalist movement as an interlocutor and the fact that more opportunities to participate were offered, as was the case in the Netherlands and in other, more consensus-based nation states, led to strategic problems. For instance, the wide-ranging public debate provoked in advance a large amount of discord within the anti-nuclear energy movement regarding the question whether it was wise to take part in that debate. After all, the radical wing felt that participation in this 'political spectacle' implied acceptance of the existing political order. Nor had the moderate wing any great expectations, yet it still felt that 'if you remained standing on the sidelines you robbed yourself of any political influence'. Many authors have described the strategic position and the dilemmas of the environmentalist movement on the basis of the political opportunity approach and other approaches. In general, these studies show that the greater participation opportunities did in actual fact contribute towards making the choice for a moderate, participatory strategy, towards a gradually less politicised and more professional environmentalist movement, and towards the movement being recognised as an interlocutor. Arts (1998) illustrated at international forums the close interconnection between the position, the impact and the strategy of the environmentalist movement. Despite so much moderate professionalism, we are now witnessing – at the national level, and particularly the international level – that a radical wing is opting for a different strategy. The well-being of animals and biotechnology in general, and cloning technology in particular, and especially the aspect of 'globalisation' have recently led to a new wave of radicalisation, including a form of radicalisation within the environmentalist movement itself.

Despite the observed radicalisation of certain groups, the moderate, participatory strategy is clearly dominant in the Netherlands. The Dutch environmentalist movement was able to gain recognition for itself as a respected partner of the government and other, established interest groups. For instance, the environmentalist movement was immediately accepted in the then 'Centrale Raad voor de Milieuhygiëne' *(Environmental Protection Advisory Council)*, from 1974 to 1993 the most important advisory body to the government on environmental policy matters (see also Chapter 3). The environmentalist movement also had membership in a variety of other advisory and consultative bodies, both at the level of central government as well as provincial and local level. It also made constant use of the various opportunities for appeal offered in the set of participatory instruments. Apart from national environmental organisations, several local groups also earned themselves a reputation with regard to successful legal proceedings against a variety of decisions and decrees, thus earning the environmentalist movement expertise in the field of the environment in terms of content, policy and procedures, in addition to political recognition. This recognition and expertise led to the accelerated institutionalisation and professionalisation of the Dutch environmentalist movement, in which the main weapons are not radicalness but a pragmatic knowledge and verification of the issues at stake.

As stated in the foregoing, the many opportunities of participation given to the Dutch environmentalist movement help to explain its choice of a more moderate participatory strategy. However, that strategy also has its risks. The Dutch environmentalist movement has been praised both at home and abroad by friend and enemy alike for its moderate approach, its willingness to enter into consultation and its high level of expertise. Yet the radical wing of the Dutch environmental movement frequently points out the other side: attaching too much significance to consultation and expertise threatens to lead to a small power base and distancing oneself from ones own supporters. This danger is not theoretical but was seen quite clearly to be real in three recent major environmental dossiers. The construction of the 'Betuweroute', a controversial rail link for goods transport from Rotterdam to Germany, led to a demonstrable gap between the local residents' groups and the environmentalist movement. On the other hand, the IJburg case, a controversial housing project in the IJmeer near Amsterdam, and the Schiphol case, the controversial expansion of the national airport, show how difficult it has become for the environmentalist movement to mobilise its supporters. In short, there is a participation paradox: the more the environmentalist movement becomes incorporated in the consultation circuit, the more difficult it becomes for it to fulfil its role as the channel for citizen participation. This places the environmentalist movement with dilemmas of a strategic and organisational nature which are comparable with those facing the trade union movement: radicalness and the ability to mobilise are difficult to reconcile with reformism and incorporation. Political influence is connected in a most complex fashion with all these strategic choices.

8.6 Participation and the societalisation of environmental policy: from 1985 to the present day

Of note in the debates on and the opportunities for participation in the 1970s is that they relate almost exclusively to government decisions, or have the intention of improving the relationship between the government and citizens. This is typical of the attitudes and customs of the 1970s. The debate on democratisation was conducted primarily as a debate about the political institutions of the nation state, and as a debate on the gap between government and (civil) society, between government and citizen. This political context also tinged the debates and conflicts on environmental issues: protests were also aimed at the government's decision-making process, even if the issue concerned was connected with the location plans or management of a particular business. Environmental protests were seldom aimed at trade and industry itself but rather at the government, which was expected to manage the environment on everyone's behalf. Environmental policy was the responsibility of the government. Even the new instruments for participation related mainly to the transparency and accessibility of government decisions.

Since the mid 1980s, both in the Netherlands and in other European countries, changes have taken place that also have an effect on the debate on and the organisation of 'participation'. These changes in environmental policy have been discussed in detail elsewhere in this book (see Chapters 2 and 3). Of major importance for the issue of participation is that, gradually, many more actors than only the government and citizens are becoming involved in environmental policy, or that environmental policy is constantly becoming more a matter for the government, the market and civil society. As has been explained in previous chapters, some of these changes are intentional, others not. That some of these changes are intentional is evident from the gradual and deliberate change in strategy and orchestration of environmental policy. In virtually all countries are regulatory strategies gradually being supplemented with more economic and communicative strategies, in which businesses and citizens, manufacturers and consumers are enticed, rather than forced to act in a more environmentallyfriendly manner. Nevertheless, the changing roles and responsibility of government, market and citizens are also the result of a much less channelled and more comprehensive process of political modernisation (Leroy & Van Tatenhove, 2000; Van Tatenhove, Arts & Leroy, 2000). This process is far wider than environmental policy and in fact boils down to a redefinition of the relationships between the government, the market and civil society. A government that works more at arm's length, deregulation and privatisation are only a few of the manifestations of the gradual societalisation and marketisation of policy, particularly of environmental policy.

Below, we look at these recent developments in so far as they are of significance for the debate on and the practice of 'participation'. We distinguish between 'societalisation' and 'marketisation'; their consequences for 'participation' can only be mapped provisionally and tentatively: these developments are not only recent developments, they are also ongoing developments, taking place in several European countries at different speeds and along different routes. We subsequently take a look at the consequences of 'societalisation ', and then 'marketisation'.

The *societalisation* of environmental policy implies first and foremost a change in management strategy, in which direct, often top-down government initiated steering is at least supplemented with other, much more communicative forms of governance (Weale, 1992). This means that citizens and non-governmental groups cannot only act reactively, for instance by making objections and instituting appeal procedures. Extensive communication takes place with these groups from the very beginning and in the successive stages of policy preparations, planning and implementation. On a semi-experimental basis, this approach has been taken in environmental policy across the whole of Europe since the early1990s in numerous projects. Especially within the sphere of regional environmental policy, the coordination of agricultural and environmental policy, of nature conservation policy, of policies recreational activities and in various major infrastructural projects have new, far more participatory models of policy preparation been used (Pestman, 2000). The variety of mediation approaches that have been used in the debates on airport expansion, from Heathrow to Zürich and from Milan to Schiphol, is typical. Whereas in former days participation had to be enforced, the participation techniques of today are part and parcel of the toolbox brought in by the process managers for that purpose. Such debates are by no means only conducted at the local or regional level. Partly inspired through the experiences with participatory technology assessment, similar scenarios are tested for national debates on aviation, biotechnology, etc., (Joss, 1998).

Secondly, societalisation implies the involvement of actors other than the well known actors to date, and the use of decision-making mechanisms other than those used up to now. Apart from the established interest groups, which now include the environmentalist movement, the government now calls upon the non-organised citizen, local residents, local associations and groups. Today, government actors, together with actors from society (and the market, see below) are the co-producers of the environmental policy that results from their consultations. Public participation and support, co-production, public-private teamwork and consensus are the keywords of this approach, and network-like constructions are the most often used organisation forms (Glasbergen, 1995 and 1998). To this end, the government adopts a variety of different roles: the so-called facilitator, the agent representing the interests of several societal and economic interests, co-financier, and obviously as the political body that carries ultimate responsibility.

This societalisation of environmental policy has varying consequences for participation, leading to both conformity with and divergence from the 1970s. First of all it goes without saying that the numerous experiments with participatory or interactive decision-making processes had the intention of widening the involvement of a variety of non-governmental actors. In that sense they are in line with the arguments made in the 1970s. Contrary to the way things were done then, the citizens and other stakeholders of today are now involved in an earlier stage and in a more open fashion in preparing policy. And also differing from the 1970s, participation today is not only concerned with breaking open decisions that have already been taken by the government, but with discussion with all parties involved: national, regional and local government, market parties and societal groups. Different concepts, scenarios, methods and techniques are used without it always being a question of a well-considered choice from any of the methodical approaches.

Even though this changed policy arena implies other opportunities for participation, it is still the question – as it also was for the participation opportunities of the 1970s and 1980s – whether it will lead to a different balance of power. Obviously, the wording of participatory decision-making processes is based on the equality of the partners. In actual fact, the various actors have a disproportionate power base at their disposal. In several Dutch cases (Strategic Green projects and the Schiphol Interim Consultative Platform) it was established that citizens and the less professional segments of the environmentalist movement in fact have a weaker position than the well-equipped professionals of the government and established economic interest groups. In the government-market-society triangle it is evident that the government-market axis is always the most well-organised (see below: marketisation and participation). Among other things this is expressed in the disproportionate degree of access among the different actors to scientific knowledge that is of great strategic importance in this sort of consultation.

Other sources of power are also divided unequally, giving citizens and the environmentalist movement a considerably weaker power base. For that matter, as was also the case in the 1970s and 1980s, this weaker position is the reason for strategic discussion within the environmentalist movement: 'should we take part of not?'. As far as the Netherlands is concerned, while the participatory approach is the dominant one, new radical sections are also emerging in the environmentalist movement, examples in this respect being the Earth Liberation Front and the Animal Rights Movement.

Despite the demonstrable differences in the social and political context, there is another parallel with the participatory innovations of the 1970s and 1980s. As was the case with the illegitimate set of participation instruments in the 1970s, such as the wide-ranging public debate on nuclear energy in the Netherlands, the newly developed participatory instruments are also at odds with the established regulations under public law. Agreements reached between the actors involved in a process of participatory decision making must, at all costs, be put to representative bodies and will possibly have to yield to other decisions taken by these bodies. Consensus at the end of an open, participatory process will therefore need to be confirmed or contradicted by the regular political bodies in accordance with the legislation in force. Therefore participation gives rise to frustration even today if (the primacy of) politics leads to a different decision than the participants fought to achieve by putting in a great deal of energy and by being willing to consult. In the 1990s the efforts for more political participation were also at odds with the demands made by the representative democracy.

For that matter, with the process of societalisation that occurred in the 1990s there was not only talk of a tendency towards more participation. On the contrary, around 1990, at least in the Netherlands, there was a part of politics that held the opinion that participation had gone too far. Especially the almost endless circle of objection and appeal procedures was giving citizens and the environmentalist movement too much power to interfere, resulting in despairing delays in the decision-making process for certain dossiers. Participation would lead to consistency and moreover would actually affect the primacy of politics. Marketisation played its role in the background in this respect too: the government wished to place as little as possible in the way of either itself, as a dynamic market party, or the enthusiasm of private businesses to invest, especially in the dossiers concerning logistics and infrastructure (the locations of waste processing plants, large-scale plans for the road, rail and air traffic infrastructure). A society that was given too much opportunity to participate stood in the way of the market in this respect. Furthermore, some people compared the situation in the Netherlands with that in other countries where, for instance, a considerably less amount of participation was tolerated regarding the decision-making process on matters of the infrastructure (as is the case for instance in the rigid top-down decision making on the TGV in France). All these considerations were part of the reason for the Dutch Government to introduce the so-called NIMBY legislation and Routing legislation. The bottom line of both laws was the combination and streamlining of opportunities to comment, whereby the circle of objection and appeal procedures was to a certain extent checked. A similar 'streamlining' of participation, coupled to the combination and acceleration of decision making, also took place in the Westerschelde Permits Act and – after the near flood disaster of 1995 – in the Delta Act on major rivers.

Apart from the political question whether this reversal of participation is a good thing, it is very much open to debate whether, firstly, the participation of citizens is the main delaying factor in such lengthy processes of decision making, and secondly, whether this slow decision-making process is so dramatic. As far as the first aspect is concerned: studies have time and again illustrated that administrative disagreement between ministries and administrative layers is a much more significant element of delay on the progress of decision making than the powers of the citizen and the environmentalist movement to interfere. However, whether the new regulations will suppress the latter still remains to be seen. Regarding the second aspect: the decision-making processes concerning nuclear energy and the Waddenzee for instance, illustrate that a lengthy decision-making process often leads to new insights and collective learning processes, and thus often results in a better decision-making process as far as the environment is concerned.

As stated earlier, other people feel that the extensive set of instruments for participation is still inadequate. Experiments are therefore carried out enthusiastically on new forms of participation, consultation and mediation. Nevertheless, it is still quite remarkable to note that the words 'political participation', as used in the 1970s, are hardly used today. 'Participatory decision making', 'public support' and 'co-production' are *the* keywords of today. These concepts are thus symbolic of political innovation in the 1990s.

8.7 Participation and the marketisation of environmental policy

We have seen a process of the *marketisation* of environmental policy running parallel with the process of societalisation since about 1985. This marketisation stands for a variety of connected indicators. In the first place, as has already been stated, there is a shift in governance style, management strategy and the set of policy instruments used. Instead of resorting to direct regulation alone, the government also introduced market-conform instruments to entice citizens and businesses to change their ways in terms of the environment. By introducing pricing mechanisms, citizens and businesses are no longer addressed exclusively as 'legal subordinates', but also as consumers and manufacturers. As parties in the market, they have their own roles to play and their own responsibilities within the operative market mechanisms.

Secondly, marketisation refers to the government leaving certain responsibilities and competencies over to the market parties (or: returning them to them). This can be done quite conspicuously through liberation and privatisation. Since the mid-1980s this has been done partly *con amore* and partly *à contrecoeur* in different European countries in environmentally-important sectors such as energy, waste, water purification, drinking water supply, traffic and communication. Privatisation and liberalisation are striking phenomena which are often politically and socially controversial, but which at any rate imply a different division of roles between the government and the market. Yet marketisation can also be less spectacular: the 'target group policy' developed by the Dutch and other national governments in the years after 1990 – now a recognised policy strategy at European level too – also implies a shift in responsibilities. Economic sectors posing a serious threat to the environment are now being called to use their own abilities and thus contribute towards solving the problems they cause. Trade organisations are also expected to enter into agreements with the government on behalf of their members, and partly to see to the implementation and monitoring thereof themselves. Spectacular or cautious, in both cases marketisation means that the role and responsibility of the government is partly expanded and partly taken over by other actors, in this case the market parties, and that market forces (in part) take the place of government regulation.

The actual consequences of privatisation and liberalisation have hardly been studied on their economic benefits and significance for environmental policy. Yet it is quite clear that they are linked to a change in the positions and responsibilities of different actors, and thus also to a change in their mutual balances of power. Their impact on 'participation' can for the time being only be pointed out intuitively. In any case, marketisation means that the citizen can no longer bring up his demands for participation solely as a *homo politicus,* but now, in principle, also as *homo economicus* or as a consumer. How waste is removed and processed, how the energy and tap water delivered to our homes is produced, is to an increasing extent no longer the responsibility of politics (alone). The privatisation of, for instance, public transport, the generation and distribution of energy, the removal and processing of waste, has partly made political participation in these fields, – as it was enforced and institutionalised in the 1970s – participation 'without an object'. Other participatory instruments will therefore need to be realised for the environment conscious citizen wanting sustainably produced energy or environmentally friendly products. There is a more or less lack of these instruments, and this also applies to other forms of quality control on the services provided by privatised companies. This has led to frequent, often fierce conflicts: on the services provided by privatised public transport companies, on the investment programmes of the private companies that supply our tap water, on the verifiableness of the claim made by energy companies that they supply 'green power', etc. These conflicts between citizens, governments, private businesses and consumer associations illustrate the unease with this new situation among citizens, politicians and trade and industry. To date there have been exceptionally few organisational instruments that can be used to guarantee the citizen-consumer access to and influence upon a 'marketised' environmental policy. Whereas today's market has increasingly become a political area because of government at arm's length, participation on that market has hardly been regulated. Neither the institutionalised environmentalist movement nor the consumer movement is apparently able to come up with a befitting strategic answer to this problem as yet. Despite the increasing supply of products claiming low energy consumption and ecological soundness, etc., and despite the increasing demand for more environmentally friendly products, there is little collective representation of interests to be seen, let alone consumer 'market participation'. At most, the individual consumer has a few sorts of eco-labelling at his or her disposal. However, on the European market alone there are many systems in operation, diverse systems of eco-labelling for a wide range of products or product categories. Partly because of this, the transparency and verifiableness of the claims made by the manufacturers in terms of environmental quality by way of these labels is virtually nil.

Also the 'cautious' marketisation, as pursued by the Netherlands in its target group policy, has implications for participation opportunities. In brief, the target group policy is as follows: the general objectives of environmental policy established by the government in its National Environmental Policy Plan are translated in consultation with the target groups into tasks specifically for the target group concerned. The environmental objective is then set out in agreements at branch of industry level and subsequently broken down into tasks for individual companies. The latter is given shape by virtue of a company's internal environmental plan and/or a permit covering the main points. The essence of the thus changed roles and division of tasks is that the government formulates the goals, and the industrial sector concerned indicates the means and in actual fact determines the speed of implementation. In the 'former' situation, the government established an overall standard and/or a standard for the entire sector, which was then basically translated in the form of an individual permit. The disposition concerned – as an achievement of the arguments for participation put forward in the 1970s – was open to appeals made by individual citizens and the environmental movement. This is still the case. And yet if we compare the 'old' with the 'new' situation from the perspective of participation we are able to detect some quite substantial changes. On the one hand, participation of the target group concerned has unmistakably increased, been formalised and regulated. Comments made by the various branches of industry on the target group policy inevitably starts and ends with an expression of satisfaction regarding 'the willingness of the government to listen and enter into consultation'. Whether this surplus of participation also leads to better acceptance cannot be answered unequivocally because of the lack of relevant empirical research. From the research into target group policy carried out to date we see in the first place that the mobilising, organising and disciplinary capacity of the branch of industry, i.e. the market parties, is of importance if target group policy is to be successful. Secondly, research on, for example, the long-term agreement on energy, shows that consultation remains restricted to a closed circle of stakeholders. While it is true that this does raise acceptance among this group, the level of participation is still limited to those directly concerned. In comparison with the 'old situation', target group policy therefore shows a strengthening of the government-trade and industry axis. On the other hand, participation among citizens and the environmentalist movement is only limited. The role of the citizen and the environmentalist movement is definitely not all that great in utilising instruments under private law, such as agreements. This public-private partnership gives access to market parties, but to a much lesser extent access to societal groups as partners. This applies at branch of industry level where the government, usually without the involvement of others, enters into agreements with target groups. It also applies at the level of individual businesses, where the authorities are gradually restricting themselves to setting out the environmental objectives in a so-called 'permit covering the main points'.

The 'how' regarding corporate environmental policy is established in an environmental management system that the company itself is responsible for. While citizens and environmentalist groups still have at their disposal instruments governed by public law, of lodging objections and appeal procedures against permits, they have in actual fact been forced into a weaker position because of the self-management given to companies.

8.8 The environment, participation and power: between the 'green polder model' and further democratisation

In thirty years of environmental policy it would seem that ' environment' and 'participation' are indeed inextricably linked. Yet at the same time, the meaning given to 'participation' and how that participation is shaped, has undergone substantial change. As part of a wider, radical political movement, the environmentalist movement has enforced participation since the 1970s with a great deal of protest and pressure. That served a dual purpose: to ensure that the environmental aspects were given more consideration in the decision-making process on the one hand, and to make governmental decisions in the field of the environment more accessible and more verifiable for the citizen and the environmentalist movement on the other. Certain aspects have led to the emergence of extensive legal and organisational instruments for participation. Participation was institutionalised along with the institutionalisation of Dutch environmental policy.

Gradual, but nevertheless radical changes have taken place in environmental policy since the mid 1980s. These changes related to more involvement being given to actors other than the government in the formulation and implementation of environmental policy. Market parties and NGOs are invited to carry co-responsibility for the development and implementation of environmental policy. In this changed context, participation is no longer motivated by the political democratic ideals of the early 1970s. Public support and acceptance are now the main arguments behind various experiments. Nevertheless, no satisfactory instruments for participation in environmental policy in the context of societalisation and marketisation have been developed as yet.

In short, the arguments for and the organisation of participation have changed considerably in the thirty years of environmental policy. Even so, this has not led to any significant changes in the balance of power in connection with environmental policy. As studies conducted in the 1970s and 1980s have shown, the citizen's actual influence on environmental policy was only slight, despite the new participation instruments, it would seem that the initiatives for participatory decision-making processes in the 1990s also failed to lead to changes in the balance of power.

Instruments for participation should not only be judged on the basis of the contribution they make towards shifts in the balance of power, to more democratisation and other exalted ideals. Those changes must also be regarded as an attempt to channel invariably new societal and political needs by renewing, time and again, the political infrastructure and the political instruments. For instance, the political infrastructure in the 1970s was evidently inadequate to resolve the often fierce environmental conflicts. Participation on the one hand and the wide-ranging public debate on the other, were 'adequate' instruments in this respect. Yet that very same set of participatory instruments has evidently been inadequate for the shifts that have taken place between the government, the market and society since the 1980s. These changed relations call yet again for their own participatory instruments.

There are two paths that can be taken to design the necessary new participatory instruments: the 'green polder model' or further democratisation, particularly of the relations between market and society. The success of the socio-economic polder model, based on a basic consensus between government, trade and industry and the trade union movement, inspires some people to put forward a similar type of model for environmental policy (see Chapter 9). Such an instrument would – in the opinion of some – substantially increase the ability to solve the problem and could settle the conflicts concerning the infrastructure, agriculture, nature, etc.

Others feel that a 'green polder model' would be no more than a disaster: they associate the 'polder model' with consultation between the elite of interest groups who, unhindered by their supporters, reach compromises with one another. Earlier in this chapter it was argued that such consensus among the elite led to non-participation. In its place, the argument is put forward for a radical step to be taken in political innovation: in the same way as society broke open the government's decision-making processes in the 1960s and 1970s, now the decision-making processes of the market parties must be made more transparent and more accessible. In other words: if trade and industry truly wishes to accept the responsibility for environmental policy, then it must justify itself in this respect. What Greenpeace achieved with regard to Shell's Brent Spar should be the logical behaviour of businesses in the future. Eco-labels, environmental reports, green energy and other instruments are then only the tentative initial steps in a process of far-reaching politicisation and democratisation of the market and society.

References

Arts, B. (1998) *The Political Influence of Global NGOs,* International Books, Utrecht.

Blowers, A. (1984) *Something in the Air. Corporate Power and the Environment,* Harper and Row, London.

Blowers, A., D. Lowry & B.D. Solomon (1991) *The International Politics of Nuclear Waste,* MacMillan, London.

Castells, M. (1974) *Monopolville; l'entreprise, l'état, l'urbain,* Editions Mouton et Cie, Paris.

Gladwin, T.N. (1980) 'Patterns of Environmental Conflict over Industrial Facilities in the United States 1970-1978', *Natural Resources Journal,* 20, 2, pp. 243-274.

Glasbergen, P (ed.) (1995) *Managing Environmental Disputes. Network Management as an Alternative,* Kluwer Academic Publishers, Dordrecht.

Glasbergen, P. (ed.) (1998) *Co-operative Environmental Governance. Public-Private Agreements as a Policy Strategy,* Kluwer Academic Publishers, Dordrecht.

Gould K.A., A. Schnaiberg & A.S. Weinberg (1996) *Local Environmental Struggles,* Cambridge University Press, Cambridge/New York.

Huberts, L. (1988) *De politieke invloed van protest en pressie: besluitvormingsprocessen over rijkswegen,* DSWO Press, Leiden.

Jamison, A., R. Eyerman & J. Cramer (1991) *The Making of the New Environmental Consciousness: A Comparative Study of the Environmental Movements in Sweden, Denmark and the Netherlands,* Edinburg University Press, Edinburg.

Joss, S. (1998) *The Role of Participation in Institutionalised Technology Assessment,* Imperial College, University of London, London.

Kitschelt, H.P. (1986) 'Political Opportunity Structures and Political Protest: Anti-nuclear Movements in Four Democracies', *British Journal of Political Science,* 16, pp. 57-85.

Kriesi, H. et al. (1992) 'New Social Movements and Political Opportunities in Western Europe', *European Journal of Political Research,* 22, pp. 219-244.

Leroy P. & J. van Tatenhove (2000) 'Political Modernisation Theory and Environmental Politics', in: G. Spaargaren, A.P.J. Mol & F.H. Buttel (eds.) *Environment and Global Modernity,* Sage Publications, London, pp. 187-208.

Lowe P.D. & J.M. Goyder (1983) *Environmental Groups in Politics,* Allan & Unwin, London.

Nauta J. & A. Steenbrink (eds.) (1985) *Het kernenergieconflict en het vraagstuk van de politieke legitimiteit,* Amsterdam.

Pestman, P. (2000) 'Dutch Infrastructure Policies in the 1990s: Changing and Contradictory Policy Arrangements', in: J. van Tatenhove, B. Arts & P. Leroy (eds.) (2000) *Political Modernisation and the Environment. The Renewal of Environmental Policy Arrangements,* Kluwer Academic Publishers, Dordrecht, pp. 71-96.

Van Tatenhove, J., B. Arts & P. Leroy (eds.) (2000) *Political Modernisation and the Environment. The Renewal of Environmental Policy Arrangements,* Kluwer Academic Publishers, Dordrecht.

Weale, A. (1992) *The New Politics of Pollution,* Manchester University Press, Manchester.

Chapter 9

GREENING POLITICAL INSTITUTIONS

Pieter Glasbergen and Henri J.M. Goverde

9.1 Introduction

One of the main themes in the debates on environmental policy relates to the question: how to intensify the problem-solving potential by capacity building? The concept of 'capacity building' in the wider sense relates to the ability to realise an organisational, procedural and normative framework within which work on environmental issues can be undertaken productively. The notion of capacity building has been stressed in OECD publications in particular (1994, p. 9). Environmental issues are seen as relatively new challenges for society, the awareness of which has developed much faster than the institutional ability to solve them (OECD, 1994, p. 11). Capacity building also forms the core of Agenda 21, the international programme of the UNCED Conference (Selman, 1999, pp. 155-156).

Jänicke sees capacity building as a combination of requirements to pursue a successful policy. He also distinguishes between cognitive-informational, politico-institutional and economic-technological conditions (Jänicke, 1997, p. 11). Cognitive-informational conditions relate to the conditions under which knowledge is acquired, disseminated, interpreted and applied. Politico-institutional conditions relate to participatory structures (openness in the policy-making process), integrative abilities (the extent of cooperation and consensus) and – as the highest form – the ability to act strategically on the basis of a long-term perspective. The third condition he distinguishes is related more to the context within which these conditions are the object of debate. As demonstrated in various studies,

P.P.J. Driessen and P. Glasbergen (eds.), Greening Society, 185–204.

economic and technological development makes a positive contribution towards a more mature environmental policy.

Also, Nelissen et al. (2000, pp. 22-23) distinguish capacity building as 'indexed capacity' and 'effective capacity'. Indexed capacity relates to the potential possibilities of (new) politico-institutional forms for the purpose of solving socially acknowledged environmental issues. The extent to which this indexed capacity is used depends on the actions undertaken by the parties concerned as inspired by their status, their values, standards and interests. Capacity building interpreted as effective capacity relates to the extent to which, and the quality with which (new) politico-institutional forms lead to a better method of approach and to the solving of those symptoms that are recognised as social issues in the contact between politicians, civil servants, citizens, businesses and experts.

In this chapter we deal with capacity building as indexed capacity and investigate the politico-institutional requirements as identified by Jänicke. We focus in particular on the debate on structuring the societal decision-making process and the rules used to achieve this. Capacity building as indexed capacity is placed in the wider issue of shaping the tense relationship between the environment and politics. After making several observations on this aspect we move on to focus our attention on the new developments observed in the institutional hierarchy in the Netherlands for tackling environmental issues. Our analysis is made on the basis of three tracks. The first track is that of political philosophy, where a debate is underway on the ethical bases of environmental politics. The second track takes us to the, often fierce, debate on the 'greening of democracy'. The third track (which more or less ensues from the first and second tracks) relates to the search for an institutional form of a more environmentally responsible development of society; in the Netherlands this is referred to as the 'green polder model'. The premises of this model and the debate on its potential value are pivotal in the following analysis.

9.2 The tension between the environment and politics

Many people are concerned about the future of the environment as the basis for the existence of mankind, animals and plants, both now and in the future. And yet values other than that of the environment often prevail in politics. Environmental issues constantly confront 'politics' (officials and institutions) with the problem of having to make choices: to reinforce the infrastructure or to limit mobility; to economically develop geographical areas or to preserve biodiversity; to minimise pollution or to create more employment; a higher consumption of energy or a lower risk of climate change. On the one hand politicians are evidently prepared to honour the environmental good on the understanding that there is sufficient support in society to do just that. The lobbyists of environmentalist groups and action groups, as well as official bodies and experts, are constantly activating people in this sense. On the other hand, politicians are wary of imposing environmental measures, which are too strict. They have no wish to be held responsible for standing in the way of free enterprise in the production, trading and transport of goods and services, and thus endanger employment. Fuelled by what is also a permanent lobby of companies, agricultural interest groups and motorists, they fear that far-reaching environmental measures will be presented to them on a platter at future elections. We feel that 'support' is an even more crucial factor than 'power', which by the way is one of the conditions for turning goals into policy. The essence of politics is to focus on discontent in society and to seek possibilities for the wide legitimisation of potential policy measures.

Policy processes in western democracies are chiefly processes of argumentation, in which being in control of orientational and organisational resources is frequently the main source of power. This is made manifest in pamphlets, debates, radio and televised reports, documentaries, symposiums, scientific studies, counterexpertise and peaceful demonstrations. Acts of violence, which confront major societal organisations with an established fact, have been pushed to the background in modern liberal democratic societies and are regarded as counterproductive. The justification of one's own, as opposed to the other person's behaviour is central; politicians to account for themselves in public use the process of legitimisation constantly. They do this in a continuous discourse with the important stakeholders concerned. They remove the tension between the environment and politics by managing various roles in their mutual connection: political leader, manager of the administrative machinery, ambassador of the public cause and facilitator in communication. The success thereof is to a large extent independent of the party political colour (Schouw & Tops, 1998; Goverde & Nooijen, 1996). 'Ecological modernisation' and the continuation of green political thought is therefore not a matter of which political party wins the elections, but rather whether society is prepared to allow the scope for a shift away from the ecological trends. Political leaders, democratically legitimised after general elections, can then – as brokers in opinions and interests – contribute towards bringing the ecological dimension into line with policy as a whole. The opportunity to do this apparently comes in waves. At the right times, the politicians must have prepared themselves in a variety of roles for using that particular policy window which offers opportunities to accelerate sustainable development.

9.3 The ethical bases of environmental politics

The tension between the environment and politics gives rise to the question of the bases of a variety of perceptions on the institutionalisation of the environmental interest. To answer this question we have a need for knowledge about the philosophical and political-theoretical perspectives on the environment, and what the significance is thereof in concrete political action. We see this as the first, most abstract search for adequate politico-institutional requirements for the greening of society. The second, parallel search relates to the demand for ideas regarding the practical options for innovation in political institutions so that they are geared more towards sustainable development (see section 9.4). The tension between the environment and politics is approached in a more normative fashion in both searches.

Nature and the environment represent a wide variety of values that have different influences on political actions. These values are underpinned in three traditions: the Christian (which includes two views), the utilitarian and the holistic (Connelly & Smith, 1999, pp. 10-18). The first, the Christian view, presumes that nature is at mankind's disposition. Nature is there to be exploited. Modern science serves as the source of knowledge and can be used to acquire power and control over nature. In fact this is a strongly anthropocentric approach. God may have created the world, but He does not live here. This furnishes ideal conditions for the natural sciences and the associated technology, which is continually becoming more dominated over nature. According to Zygmunt Bauman (Munters, 1998, pp. 48 and 56-58) this line of argument forms the basis for modernism, which is allegedly the main cause of the risk society in which we live today (Beck, 1992). The alternative Christian view envisages mankind's obligations towards nature. Man is the administrator, and God has given the Earth to him on loan. This view obviously leads to man taking care of the environment. This alternative view is therefore more ecocentrically oriented. Property rights don't have an absolute value in this view, and nature is seen not only as a vehicle for human goals. People are not expected to acquire their property at the expense of future generations.

Conversely, utilitarians judge man's deeds primarily on the basis of the consequences, not on their propriety in the moral sense. The ultimate criterion is the realisation of 'the greatest good for the greatest number'. This can imply that society should assign scarce resources to those who are already better off on the understanding that it makes a contribution towards maximising the total satisfaction within a community. This criterion of societal satisfaction deserves a comment. If the accent comes to lie too one-sidedly on the success of the process, then there is an increased likelihood that the result in terms of content – whether the environment will benefit –

carries too little weight. Furthermore, by placing too much emphasis on a good outcome of the process it is quite possible that certain interests are excluded (e.g. the interests of minorities or future generations). This would therefore not be seen as a satisfactory solution to an environmental problem in a democratic constitutional state. Another question is also relevant here: 'Who counts when we are talking about the greatest number'? Only people, or also (parts of) the non-human world (flora, fauna and possibly minerals)? In brief, it would seem that utilitarianism devotes less care and attention to certain moral institutions such as the value of life, the value of ecological systems, the existence of species and the more anthropocentric ideas on justice.

It is these institutions that are given the main place in the third tradition, that of holism. The point of departure here being that all life is correlated, and as such must be respected. Ecocentrism is more in the foreground here too. This school stresses the continuity of mankind within the whole of the natural world. In an extreme form, this belief can be found in the 'Gaia hypothesis' (Lovelock, 1979), in which the world is seen as a single, totally functioning, living, self-regulating biosystem.

These three traditions widely differ, and by virtue of their multiformity it is impossible to draw up an unambiguous view on political intervention in environmental issues. Conflicts in the political arena as to the 'right' approach taken on environmental issues are consequently unavoidable. At the same time it is only right to draw the conclusion that there are more lines of argument that ensue from the above traditions, each of which have a more or less demonstrable effect on the actual political positions taken by the different political parties, environmentalist organisation and environmental experts. Some devote a great deal of attention to the future of the next generations. Others feel it unjustifiable to subordinate the needs and interests of the present generation. The worldwide acceptance of the concept of 'sustainable development', as a normative guideline for political intervention, has not counterbalanced the predominant value conflicts. Yet the concept has been advantageous in finding a new basis for discussion to reach practical solutions. The broad formulation, giving attention to environmental protection as well as to social and economic development, has made the traditional antagonists, such as environmentalist organisations and trade and industry, interdependent players in one and the same problem area. Entrepreneurs in professionalised environmentalist organisations and enlightened entrepreneurs are able to unite in new partnerships (Glasbergen & Groenenberg, 2001). The political world is more than pleased to associate with this consensus. Practical environmental politics has therefore entered into a moderate utilitarian discourse. Thoughts on the desired political institutions for the purpose of achieving an ecological breakthrough are consequently fuelled less by the increasingly more marginal Christian and holistic tradition.

9.4 The ecological restructuring of democracy

The 'green parties' have meanwhile come into power in the governments of several nations (particularly in Germany, Belgium and France). The process of ecological power building has therefore certainly been demonstrably successful. Yet the formal power of the green parties is still incongruent with the strength of their views. There are two standard, but contradicting answers to this heuristic from the ecological point of view regarding the relation of green politics and democracy.

Some claim that green political parties should exert pressure to achieve a more participatory democracy (see also Chapter 8). Liberal democracy – based on representation and therefore indirect – is supposedly inadequately responsive to govern the citizens' behaviour on the basis of ecological priorities. Both the material inequalities among citizens, as well as the bureaucracy-dominated political organisation, would prevent the emergence of new forms of public space in which the self-interest of individual consumers is exceeded. In addition to participatory forms of political organisation, a radical institutional decentralisation would be necessary (both in the field of production and the field of administration) to reach an ecologically justified lifestyle. The 'green/autonomous' political powers organised via the Internet against the increasing dominance of the 'institutions of globalising capitalism' (the World Bank, IMF and WTO) such as those that became manifest in Seattle (1999), Prague (2000) and Genoa (2001) are recent, somewhat extreme examples of this trend (Smith & Smythe, 2000).

Others choose to take the point of view that democracy should rather be seen as an obstacle standing in the way of solving the ecological crisis. Only a strong government will be capable of limiting the freedom of citizens to prevent further ecological degradation. Individuals, who are concerned with their own interests only, even if they threaten the future population, need a government, which is able to act strongly in the general interest. While the first viewpoint implies libertarian-socialist sympathies, the second can be reduced to anti-democratic, autocratic arguments.

However, these extremes in green politics have recently been worked out in more moderate terms. The more support environmentalist organisation and green politics gain, the more the emphasis in actions must be shifted from raising problems (setting up the agenda) to contributing to solutions (developing alternatives and process management). Many 'greens' have been engaged in this since the 1980s to make sure that their ideas are honoured by forming new alliances with established politicians and citizens. That they have been successful is apparent from the greening of the traditional political parties, which are increasingly incorporating the interests of the environment into their own political programmes. It goes without saying that this approach taken by the traditional parties also works as a powerful strategic impetus against too strong a green movement. In fact a joint interest is increasingly attached to the search to find out how green politics are able to make an analytical and normative contribution towards innovating democracy. This search focuses on three issues (Doherty & De Geus, 1996, pp. 3-15):

- The discourse of the green movements on democracy. a) Priority given to decentralisation, in which there is an internal battle about the approach taken, either through the 'bio regions' (the size being determined by the sustainability of the community with the lowest disturbance in the ecological balance) *or* through the dominance of non-territorial 'green political communities'. b) The argument for participatory democracy aimed against the power of bureaucracy and against inflexible planning systems, including debate on the boundaries of civil disobedience (when will non-violence become inappropriate?) and democratisation, also outside the political sphere (work, family, school, church).
- The status of democracy in the way of thinking in green politics. Does the nature of environmental issues enforce a redefinition of the concept of 'political community' (transnationality; obligations to future generations)? Must the 'moral community' be expanded to incorporate the inclusion of the intrinsic value of nature, and what does this mean for democracy as an appropriate form of political organisation?
- The political institutions able to establish a green democracy (a problem of legitimacy; how can the 'greens' become effective democrats? does the concept of 'eco-state' offer a solution in this respect? what are the ecological rights and obligations, and how do they relate to the majority decisions in a representative democracy?).

This search is by no means at an end as yet. The direction can, however, be set out: an atmosphere of political realism among the 'greens', in which a substantial part of the 'green discourses' is incorporated into the fundamental starting points of liberal politics and the democratic, constitutional state.

9.5 The polder model as a source of inspiration

In the foregoing we have seen that:
- Politicians are starting to act more as politically-neutral managers;
- The concept of sustainable development fuels moderate utilitarianism;
- The greens are more likely to embrace political realism.

These trends form the breeding ground for a third, more practically oriented search towards practising environmental politics. This reorientation came into being in the Netherlands in the 1990s under the influence of the many societal conflicts on the environmental effects of a large number of newly planned infrastructural works (roads, railway track, ports, airports, etc.), resulting in a debate on the 'green polder model'. The point of reference for this debate was the experiences with institutional structures for settling conflicts in socio-economic policy.

The term 'polder model' refers to one of the Dutch elements that determine its culture: the 17th Century polders that offered protection from flooding and opened up new agricultural land. These polders could only be realised through cooperation and consultation, focusing on the profitable investment of substantial, often foreign, flows of capital from private individuals. The concept was adjusted at a later date on specific characteristics of socio-economic policy and the institutions in the Netherlands and in this sense – especially in the second half of the 1990s – have attracted international attention.

The metaphor of the 'polder model' in the Netherlands refers to the combination of the predominant culture and institutions. According to Delsen (2000, p. 169) Dutch culture is an individualistic, consultation-oriented culture in which immense value is attached to the aspects of solidarity and equality. Intercultural research (Hofstede, 1982) has shown that the Netherlands is one of the most 'feminine' countries of the world: the emphasis being on the search for collaboration and consensus between heterogeneous stakeholders and on an equal distribution of income and power. Hampden-Turner & Trompenaars (1994) pointed out that Dutch people prefer to negotiate and meet other people halfway, even if it is at the expense of their own cultural values, yet at the same time they remain analytical. The main characteristics of the polder model are the organised solidarity and organised decentralisation.

In the main, the Netherlands fits well into the so-called 'Rijnland consultation model' or the consensus model. Deciding by majority rule, the Anglo-Saxon model is far distanced from the Dutch system (Lijphart, 1999; 1976). The essential political institutions are the institutionalised consultation and consensus forming among the representatives of employers' organisations and the unions at central level. Terms of employment are drawn up collectively by employers and workers for entire branches of industry, which the government then declares legally binding for non-union members (Collective Labour Agreements). Also typical are the strict rules that offer protection from dismissal, the extremely complex system of social security, and the mature welfare state (Delsen, 2000, p. 169). As soon as the institutions are no longer of any use to society, i.e. when no one feels it appropriate to keep them in existence, they are adapted (Muyskens, 1998). A clear example of this mechanism was the discontinuation of the permanent negotiation council between employers and workers in the agricultural sector (Board for Agriculture) and the Minister of Agriculture in 1995. This exponent of the neo-corporatist structure (the bastion of the so-called 'green front') fell definitively when it appeared that none of the participants were prepared to defend it (see Chapter 6). It was also proven here that the sustainability of the polder model is not only dependent upon the domestic socio-economic relationships but also on the international economic and political dynamism in particular (in this case the establishment of the WTO after the adoption of the Uruguay round and the essential changes to the Common Agricultural Policy of the EU; Goverde, 1999, 2000).

The success of socio-economic policy pursued in the Netherlands has been referred to as 'a Dutch miracle' (Visser & Hemelrijck, 1997). The continuing pay restraint policy was often said to be responsible for this (Delsen & De Jong, 1998), but was also disputed in these effects (Becker, 2000). However, the policy dimension of the polder model does in fact refer to an integrated policy of pay restraint, an easing of the tax and premium burden and a reduction in collective expenditure, deregulation, and the linkage of national currency to the German Mark (Delsen, 2000, p. 10). As such, Watson et al. (1999) point out that Dutch policy is not so much a miracle but rather a standard solution that is to be found in all textbooks on economics. The Dutch situation illustrates exactly how reciprocally stimulating the influence of relative orthodox economic instruments, both macro-economic and structural reforms, can be, especially if they have the positive, wide support of society. An important task of the Dutch institutions is to acquire that support, although the intensity of that task should not be exaggerated. After all, the same institutions were responsible for causing a substantial delay in the 1970s. One remarkable aspect of all this is that the stakeholders in the socio-economic field (the social organisations of employers and workers, political parties, and often businesses as well as interest groups and pressure groups) accept the results of the economic models of the Central Planning Office as the point of departure for their negotiations. The economic profession therefore does a great deal in establishing the rules of the polder model. Nevertheless, of decisive significance for the success of Dutch social economy in the 1990s would appear to be the predominant culture and the analytical ability to realise the necessity and to envisage the course taken to adapt – what in themselves are unique – institutions to the changing international and global environment.

9.6 The 'polder model', a source of inspiration for new institutions in environmental politics

The socio-economic 'polder model' offers such a successful policy and institutional framework that it is only obvious that, on the subject of environmental issues, stakeholders ask themselves – for instance if the issue concerns the effects of major infrastructural work – whether this model can also be used in the decision-making process for the physical environment, particularly environmental policy. However, before we declare the polder model sacrosanct for environmental issues it is important to realise that an inherent paradox is concealed in this socio-economic polder model. After all, the greater the success of the 'polder model', the less the individual parties stand out on the national political platform. The contribution made by the parties is less recognisable. This gives rise to problems in setting out the individual parties' distinctive features for their supporters. Successful collaboration consequently leads to a certain undermining of the consensus model and therefore also to problems of legitimisation. The polder model can thus be ruined by its own success. Nevertheless, the socio-economic decision-making process in the Netherlands has gradually become the model for a way of practising politics in which opposite interest groups try to reach agreement on the starting points of policy. If necessary, the government acts as the mediator, yet this role is still mainly one of sanctioning the outcomes of consultation. This is widely agreed to as a theory for political actions. While this theory has neo-corporatist features it is essentially different in the sense that the privacy and the elite character of the neo-corporatist system are lacking. The advocates of NGOs play an important role, and there is also scope for (moderate) conflict. Moreover, politics fulfil a crucial role ('the primacy of politics'). In other words, the polder model shows a large degree of conformity with what Lafferty & Meadowcroft refer to as a 'cooperative regime'. Such a regime encourages groups from a variety of social sectors, each with their own perspectives and interests, to develop a common policy-making method for which they also take joint responsibility (Lafferty & Meadowcroft, 1996; Meadowcroft, 1998). Adapted from Weggeman & De Jong (2000, pp. 88-90) we can set out three points of departure for these new institutions for environmental politics:

- *A pluralist view on public space.*
  Environmental policy is not an area that can be controlled by a national government from a central position, but one in which many

government organisations (local, regional, functional) and non-governmental organisations fulfil a constituent role. They are recognised as stakeholders.

- *An institutional recognition of differences.*
  Values and interests can differ in the political arena where environmental matters are at issue. This is not an obstacle but a condition for the effective application of rules, and thus for the integral solving of problems. It is recognised that there is a diversity of insights, which can lead to consensus through an exchange of interests.
- *The acceptance of self-regulation within certain frameworks.*
  In the public space, a certain amount of authority is assigned to parties other than the national government, to take upon themselves a part of the decision-making process and the subsequent implementation in their own circles.

9.7 Towards environmental and political innovation on four institutional levels

Although there is a high level of agreement among the public as to the goals of environmental policy – established for the long term and quantified in national environmental policy plans – the number of conflicts surrounding the planned physical space interventions increased considerably in the 1990s. Examples in this respect are: the Rotterdam Port expansion, a new rail link for freight transport to Germany, a new rapid rail link connecting up with tracks in Belgium and France, the expansion of the national airport Schiphol, and the realisation of a principal structure for nature development. Environmentalist organisations were always informally involved in the development of these plans. A new national policy document on spatial planning has also been drawn up in a broadly based participatory process with the NGOs. However, the existing consensus on the long-term goals in the field of the environment and spatial planning is unable to prevent the occasional fierce battles that flare up among those concerned during the decision-making process on specific projects. Again and again do we see the lack of public support for the projects, and the decision-making process becomes entangled in a net of legal proceedings instituted by local, regional and national (environmentalist) organisations. There is also opposition from lower government bodies such as municipal and provincial authorities. There is no lack of formal participation in these projects. Yet while the government – the initiating body – tries to make use of the participatory procedure to ensure that its plans are legitimated, the participating NGOs frequently first demand a fundamental discussion on the necessity of the new infrastructure. In practice this leads to a confusing debate as to how the problem should be worded. The debate then frequently degenerates into a debate on the purpose and necessity of the project, the precise route to be taken, and the aspect of blending in with the countryside.

These experiences were the direct reason for a more fundamental discussion on the functionality and legitimacy of the existing decision-making arrangements. The idea is that the problem definition and the development of options to solve the problem should no longer be a government-internal matter (Pestman & Van Tatenhove, 1998, p. 262). It is on this particular aspect that a link is made with the socio-economic polder model. Acting as a source of inspiration, this model leads to the development of a varied discourse on the renewal of the institutional frameworks for practising environmental politics.

The original idea, initiated by a rather small 'green' political party, was to give environmentalist organisations a place in the consultation circuits associated with national politics. One of these circuits is the institutionalised socio-economic consultation with the trade union movement and trade and industry (Social and Economic Council (SER)). Independent experts participate in this prestigious, governmental advisory body as a third party. Environmentalist organisations should be given a permanent place in that body as the fourth party – the 'green trade union movement'. Ecologically and economically oriented parties could then tackle the problems efficiently and effectively in such a body; they could achieve a win-win situation. This 'green polder model' was received with enthusiasm by professional environmentalist organisations, with the exception of Greenpeace, an organisation which sees structural consultation as a disturbing element when taking action. All other organisations see it as an opportunity to penetrate through to the core of the socio-economic decision-making process and to achieve an exchange of interests in negotiations without affecting the opportunity to implement their plans of action (compare strike action as the trade union movement's ultimate weapon).

As the main organisation in the establishment of policy, the government should take a small step down in this 'green polder model' without, however, affecting the primacy of the politicians. It would preferably be given a different form, that of formulating the goals and requirements, the frameworks within which public actors can make decisions in consultation. Four levels of decision-making are distinguished in the advocated model:

- Interest groups could give their advice on general government policy with an ecological component in a permanent consultation structure. A new 'green office' of the socio-economic consultative body could take this task upon itself.

- Holding discussions on the purpose and necessity could start up major infrastructural projects. A set structure with permanent participants is considered important in this respect too. Scope could also be created for the participation of regional and residents' groups who feel they have an interest in the project.
- Situations whereby, although the decision on the principle of a project receives wide approval, its implementation gives rise to many questions and debate (e.g. the choice of route for a new rail track). It is quite imaginable that decision-making on those issues is left up to societal organizations, while the government is responsible for formulating the frameworks; it need not take part in the negotiations. Here too is a permanent core of participatory parties thought to be desirable, supplemented with relevant interest groups.
- Agreements to which central government both are and are not a party. Environmentalist organisations are not considered a party for the major environmental agreements between government and trade organisations; they lack the necessary potential. Nevertheless, it is considered quite feasible for environmentalist organisation to enter into agreements with businesses at the forefront of the environmental field. The government could even declare such agreements binding for an entire branch of industry (Van den Biggelaar & Wams, 1999, pp. 175-177).

In this proposal we see an environmental movement, which has meanwhile become so professional that it is able to leave the informal and private negotiatory circuits to claim a place at the major, formal negotiating tables. Indeed, it has positive expectations in this respect: if the government takes the outcomes of the negotiation at these levels seriously, decisions will be able to be made far more quickly and will be able to count on wider public support; at least this is what is anticipated. The modernisation of democracy – which is considered essential – is thus able to go hand in hand with a more efficient and more effective decision-making process.

9.8 Diversity in the green polder

If we look at the proposals for the 'green polder model' in the light of the three domains set out in section 9.4, in which green politics could contribute towards the modernisation of democracy, then it concerns somewhat moderate suggestions in the second (a redefinition of the political community) and third area (the political institutions through which a green democracy can take shape). Moreover, the expectations regarding the effect of the green polder model apparently soon differ in practice. Whereas some parties seek for a fundamental modernisation of democracy by way of institutional change, the perspective taken by others is focused on a more effective decision-making process by eliminating friction, i.e. a decision-making process that does not become constantly entangled in legal proceedings.

Two variants can be distinguished in the first approach. The model set out in the foregoing is one of them. An attempt is made to integrate the environmentalist organisations into the structures for the socio-economic decision-making process. All economic activity can then be examined immediately against the ecological consequences. The expectation is that conflicts will then be easier to resolve in a permanent negotiatory situation than in an ad hoc situation. The parties are familiar with one another in a permanent consultative body, they become accustomed to one another. Moreover, they are aware that they will need one another in the future too. This increases the willingness to reach an exchange of interests (VROM-Raad, 1998, p. 96). Others feel that the inclusion of environmentalist organisations as a new party in the institutionalised socio-economic consultations also institutionalises the imminent conflict between the environment and the economy. The question is whether this does not envelop the environmental interest, and thus strengthen the established economic interests. Looked at from this point of view, maintaining this conflict is positively valued as a means of social change. The argument here is for an institutional change – not in, but analogous to – the organisation of the socio-economic decision-making process. Issues of sustainability should be dealt with by a separate Council that works alongside the Social and Economic Council. The view in this respect is that a sound market has a need for strong countervailing institutions (Nauta, 1999, p. 145). Also analogous to the socio-economic policy, such a Council could conclude Collective Environmental Agreements with the representatives of significant economic interests.

Conversely, the view in the second (pragmatic) approach is that no form of legal and permanent institutionalisation of environmental interests is required in the societal decision-making process. The argument put forward in this approach is for a more interactive form of ad hoc decision-making on concrete issues. Particularly Dutch Government holds this point of view. From this perspective, green consultation can prove to be a useful addition to the consultations between government and parliament. Environmentalist organisations may participate, be specifically invited to do so, yet the primacy of decision-making is with the politicians. A partial transfer of the decision-making process to a consultative body of societal groups is therefore out of the question. The diagnosis that there is a democratic deficit in society is rejected. The Minister for Housing, Spatial Planning and the Environment (with nods of approval from the ministers for Transport and Public Works and Economic Affairs) stated in parliament that the "democratic process between the government and parliament must remain in place" and "... that the elections have ensured adequate representation of the green element in society in the House" (NRC Handelsblad, 17 June 1999). Yet it is still accepted that a change must be brought about in the organisation of the decision-making process on major projects. Pivotal in this contribution to the discourse are considerations of efficiency. By sitting at the table together with a wide variety of non-governmental organisations, says the Minister, the Cabinet is able to form a good idea as to how society judges certain problems. The politicians can take advantage of that. Moreover, it is not out of the question that this limited green polder model will develop itself further in the organic sense.

9.9 Discussion

Whichever version of the green polder model is latched onto, the basis will still be formed by the elements of the new political theory (section 9.5) and the four institutional levels explained in the above. As an abstraction of the experiences of socio-economic policy they form a discourse on a new institutional framework for environmental policy. A constitutional power apparently emerges from this discourse.

In the most recent wave of public and political involvement in environmental issues it was generally accepted that environmental considerations deserve to be given a permanent place in the societal decision-making process on the shape of society. We can observe that the intention of the relevant NGOs is to play a greater role in the realisation, shaping and implementation of government policy, not only in the ideological sense, but also in the factual sense. There is also a simultaneous renewed process of growing conflicts of interest, especially regarding new physical infrastructural developments. The inclusion of environmental considerations is thus not a technically neutral issue, but precisely a political issue. It is clear that a new institutional structure must be found to deal with this. Society is gradually developing into an organisation of organisations. Policy can thus no longer be developed without the support and legitimisation of a wide field of organised interests. However, this view leads to fundamental issues, both for the environmentalist movement and for formal political life.

Environmentalist and nature preservation organisations have an impressive number of members in the Netherlands, a number that outruns by far the number of members of political parties. On these grounds, various organisations claim that the shaping of society cannot be left primarily up to the government or the market, but that civil society itself must become actively involved (Duyvendak, 2000). Yet the first question regarding the environmentalist organisations is whom do they really represent. This question relates to both the actual substance of the interest as well as their supporters. Environmentalist organisations often pretend to be the advocates of weak interests; they speak up on behalf of nature and future generations. This is a much wider goal than those of the social partners in socio-economic policy who choose their position on the basis of the direct material interests of employers or workers. The environmental issue, says Vis (1999, p.139), is outside a social framework in which groups that are each other's countervailing power has the say. This means that the sole right of environmentalist organisations with regard to the interests they represent can be brought into dispute. Doubts can also be cast on the legitimisation of the environmentalist movement on the basis of their membership figures. Membership is mainly passive and anonymous, based on sympathy with the abstract goals, and not so much on the promotion of interests. This implies that it is weaker in representing its supporters than are the employers' organisations and unions. All further institutionalisation will inevitably also have consequences for the organisation of the internal democracy. It should also be questioned whether the weapons – campaigns, boycotts, legal opposition – are adequate to create sufficient scope for taking action in negotiatory situations. Two recent public debates, namely on the future of Dutch civil aviation and the Port of Rotterdam, have shown that this is indeed extremely difficult. The debate on aviation could end in no other way than with the divided opinions of trade and industry and the environmentalist movement. The environmentalist movement saw no other choice than to step out of the consultation process. However, it appeared that all parties were positive about the debate in which the reciprocal arguments could be clarified. Despite the failure of the 'green polder' consultations, the benefit in this case was the consensus that Schiphol Airport should be dealt with as a regular company for which environmental limits are formulated. The debate about the seaport initially ended in a stalemate. The parties did, however, continue their discussions and agreement was recently reached between the environmentalist organisation and trade and industry. The port can be expanded on condition that the expansion is accompanied with a large-scale new nature reserve, and that the environmental aspects are taken more keenly into consideration when refurbishing the old port areas. It is impossible to eliminate fierce conflict with a few months of discussion, but the fact that dialogue has been entered into, and is said to be beneficial, is a major change. The case history set out in the foregoing, and also the departure of the agricultural organisations from an intensive public debate on the use of crop pesticides for instance, shows that open conflicts are regarded as very important in the 'green polder model'. This leads to the following observations. Firstly, these junctures reflect that the stakeholders have still not formed a definitive picture of the most desirable institutional form for the 'green polder model'. The model is thus not established as yet. Secondly, the ongoing discourse on the 'green polder model' gives each party a reason for setting out their own distinctive features. They do battle by claiming that the opponent's specific behaviour is completely out of line with the model and thus justify themselves – certainly to their own followers.

From the statements made by the Minister for Housing, Spatial Planning and the Environment we can deduce that formal politics, the representative democracy, also has difficulty with the different versions of the polder model. Both government and parliament doubt all the bases of the model. This applies with regard to the 'pluralist view on public space', for the 'institutional recognition of differences', and especially for the 'self-regulation', even though the frameworks are set by formal politics. There will always be a certain amount of tension between the views on the green polder model and the system of representative democracy. The government, in the representative democracy, cannot fully transfer its competencies on the main issues in society to the NGOs. For politics, the main question for a more interactive management is inevitably how it can do justice to the electorate's mandate. After all, the green polder model can lead to the co-responsibility of numerous persons but no one really accepting it (VROM-Raad, 1998, pp. 89-90). Also, with a wider circle of participants in the pursuit of policy, the government must guarantee the democratically legitimated procedures and protection of the rights of non-stakeholders. This, in particular, is one of the tasks the government cannot transfer to another party. Yet the government is still faced with a dilemma. On the one hand it would seem that reducing the role of government to one of a process facilitator is fundamentally contradictory to the constitutional order of the neo-liberal, democratic society and the constitutional state. On the other hand, a system of majoritarian rule, often used in the Westminster model, may not be introduced in a chiefly Rijnland consultation model for collective decision making because of the formal political scene.

9.10 Conclusions

The process of capacity building, both as indexed capacity and effective capacity, can only proceed by degrees, certainly in a model, which is mainly consensus, based. The search undertaken in the Netherlands to index the capacity such that in due course the system will perform better has been set out in the foregoing. The transformation to a 'green polder model' has taken shape in a provisional new structure of consultation. Representatives of the environmentalist movement have been taken into two of the socio-economic consultative body's committees. The government has also formulated a framework for project-based consultation with non-governmental organisations on issues of sustainability, spatial planning and the infrastructure. Departing from the primacy of politics, this extremely brief framework gives emphasis to the necessity of involving the NGOs in the opinion-forming process at an early stage. As a rule, this will be in the stage when new national policy documents are being prepared and the purpose and necessity of infrastructural projects are being considered. Consultation and the exchange of arguments are also considered important in the later stages of preparing decisions. Non-governmental organisations can themselves take the initiative to enter into consultation of this sort; the minister primarily responsible must respond to this consultation within a period of six weeks. This is the first step towards a decision-making model in which a move can be made towards reaching a final result in the various stages of the policy-making process, which inevitably involves periods of political decision-making. The decisions that round off a completed stage will serve as the framework for the start of the following stages. A constitutional consultation will take place at the start of the trajectory on the subject matter and sort of consultation. This will ensure that the system is kept highly flexible. The outcome of the consultation will also serve as input for the political decision-making process. Shaped in this way, the polder model will circumvent some of the key issues of a more institutionalised form. At the same time, NGOs will be recognised in terms of their importance and they will be activated as a source of knowledge in the political process. This could be a step towards a new institutional order. It is quite possible that opportunities will arise for unexpected coalitions between politics and NGOs and also among the NGOs themselves. Simultaneously, whether the parties involved are prepared to meet each other on an equal footing will only be seen in practice. After all, the green polder model will only function properly if the balance of power between all parties is more or less equal and there is an awareness of their interdependency, not only now but in the long term as well.

References

Beck, U. (1992) *Risk Society. Towards a New Modernity,* Sage Publications, London.

Becker, U. (2000) *Successful Adaptation by Consensus? Employment Development, Power Relations and Hegemony in the Dutch 'Delta Model',* paper IPSA, RC 36 'Political Power', Quebec.

Connelly, J. & G. Smith (1999) *Politics and the Environment. From Theory to Practice,* Routledge, London/New York.

Delsen, L & E. de Jong (eds.) (1998) *The German and Dutch Economies: Who Follows Whom?,* Physica Verlag, Heidelberg.

Delsen, L. (2000) *Exit poldermodel? Sociaal-economische ontwikkelingen in Nederland,* Van Gorcum, Assen.

Doherty, B. & M. de Geus (1996) *Democracy and Green Political Thought. Sustainability, Rights and Citizenship,* Routledge, London/New York.
Duyvendak, W. (2000) 'Het Groene Poldermodel kan niet genegeerd worden', in: *NRC Handelsblad,* 15 september.
Glasbergen, P. & R. Groenenberg (2001) 'Environmental Partnerships in Sustainable Energy', *European Environment,* vol. 11, no. 1, pp. 1-13.
Goverde, H. (1999) 'Managing Integration and Marginalisation for the New Europe', in: *Notas Economicas, Revista da Faculdade de Economia da Universidade de Coimbra,* Outubro '99, numero 13, pp. 95-106.
Goverde, H. (2000) 'EU and Agro-politics', in: H. Goverde (ed.) *Global and European Polity? Organizations, Policies, Contexts,* Ashgate, Aldershot, pp. 253-268.
Goverde, H. & M. Nooijen (1996) 'Bestuursstijl van succesvolle wethouders: politieke managers zonder partijpolitieke kleur', *Openbaar Bestuur,* nr. 9, pp.2-8.
Hampden-Turner, C. & F. Trompenaars (1994) *Zeven gezichten van het kapitalisme in de VS, Japan, Duitsland, Frankrijk, Groot-Brittannië, Zweden en Nederland,* Uitgeverij Contact, Amsterdam.
Hofstede, G. (1982) *Cultural Consequences. International Differences in Work-Related Values,* Sage Publications, California.
Jänicke, M. (1997) 'The Political System's Capacity for Environmental Policy', in: M. Jänicke & H. Weidner (eds.) *National Environmental Policies. A Comparative Study of Capacity-Building,* Springer, pp. 1-24.
Lafferty, W.M. & J. Meadowcroft (1996) 'Democracy and the Environment: Prospects for Greater Congruence', in: W.M. Lafferty & J. Meadowcroft (eds.) *Democracy and the Environment; Problems and Prospects,* Edward Elgar, pp. 256-272.
Lijphart, A. (1999) *Patterns of Democracy. Government Forms and Performance in Thirty-Six Countries,* Yale University Press, New Haven.
Lijphart, A. (1976) *Verzuiling, pacificatie en kentering in de Nederlandse politiek,* De Bussy, Amsterdam.
Lovelock, J.E. (1979) *Gaia. A New Look to Life on Earth,* Oxford University Press, Oxford.
Meadowcroft, J. (1998) 'Co-operative Regimes: A Way Forward?', in: P. Glasbergen (ed.) *Co-operative Environmental Governance; Public-Private Agreements as a Policy Strategy,* Kluwer Academic Publishers, Dordrecht, pp. 21-42.
Munters, R. (red.) (1998) *Zygmunt Bauman. Leven met veranderlijkheid en onzekerheid,* Boom, Amsterdam.
Muysken, J. (1998) 'Job Growth and Social Harmony. Reflections on the Dutch Polder Model', in: *Meteor Working Paper,* WP/98/005, Universiteit Maastricht.
Nauta, J. (1999) 'Terwille van een politiek die vooruitziet', in: W. Duyvendak, J. Horstik & B. Zageman (red.) *Het Groene Poldermodel. Consensus en conflict in de milieupolitiek,* Amsterdam, pp. 142-149.
Nelissen, N., H. Goverde & N. van Gestel (2000) *Bestuurlijk vermogen. Analyse en beoordeling van nieuwe vormen van besturen,* Coutinho, Bussum.
NRC-Handelsblad (1999) *Groen Poldermodel is nog niet veel meer dan een kreet,* 17 juni.
OECD (1994) *Capacity Development in Environment,* Paris.

Pestman, P. & J. van Tatenhove (1998) 'Reflexieve beleidsvoering voor milieu, ruimtelijke ordening en infrastructuur: Nieuwe initiatieven nader beschouwd', *Beleidswetenschap*, 12e jrg., no. 3, pp. 254-272.

Schouw, G. & P. Tops (1998) *Stijlen van besturen,* Atlas, Amsterdam.

Selman, P. (1999) 'Three Decades of Environmental Planning: What Have We Really Learned?', in: M. Kenny & J. Meadowcroft (eds.), *Planning Sustainability, Routledge,* London, pp. 148-174.

Smith, P.J. & E. Smythe (2000) *Sleepless Seattle: Challenging the WTO in a Globalizing World,* IPSA, paper, Quebec City.

Van den Biggelaar, A. & T. Wams (1999) 'Overleg met een stok achter de deur', in: W. Duyvendak, J. Horstik & B. Zageman (red.) *Het Groene Poldermodel. Consensus en conflict in de milieupolitiek,* Amsterdam, pp. 174-178.

Vis, J. (1999) 'De illusie van de groene variant', in: W. Duyvendak, J. Horstik & B. Zageman (red.) *Het Groene Poldermodel. Consensus en conflict in de milieupolitiek,* Amsterdam, pp. 135-141.

Visser, J. & A. Hemelrijck (1997) 'Lecture', in: R. Giebels, *Seminar Duitsland Instituut, Van 'Dutch disease' naar 'Dutch miracle' in tien jaar – Een verklaring voor het Nederlandse model,* http:/www.xxlink.nl/dia/diahome/giebels.html.

VROM-Raad (1998) *Advies over de sturing van een duurzame samenleving,* Advies 006, Den Haag.

Watson, C.M., B.B. Baller, J.K. Martijn & I Halikas (1999) *The Netherlands. Transforming a Market Economy,* Occasional Paper 181, International Monetary Fund, Washington DC.

Weggeman, J. & M. de Jong (2000) 'Het groene poldermodel in perspectief', *Beleidswetenschap,* nr. 1, 14e jrg., pp. 72-96.

Chapter 10

THE NETHERLANDS: INTERNATIONAL INNOVATOR OR SELF-INTERESTED NEGOTIATOR?

Bas Arts, Carel Dieperink and Duncan Liefferink

10.1 Introduction

That there is a question of a paradigm shift in Dutch environmental policy, both at the level of discourse on policy and its implementation, has been pointed out several times in this book. This paradigm shift is also referred to in this chapter as 'policy innovation'. The key words in this change include: a differentiation in the objectives, flexibility in the set of instruments, openness and interaction in implementing policy, and the integration of policy discourses (e.g. 'the environment and space', 'the environment and the economy', 'sustainable development').

Similar policy innovations are also under way at European and international level (although these are generally to a lesser extent and later than those at the national level). These parallel developments give rise to the question whether there is a link between policy innovations in the Netherlands and those in other countries. The following definition of the problem is central in this chapter:

- To what extent is the Netherlands co-responsible for policy innovations in international environmental policy, and does this make the Netherlands an international innovator? and
- if so, what are the reasons for the Netherlands to play this role?

This problem is dealt with as follows. First of all we look at the reasons behind the active environmental diplomacy in the Netherlands. We then outline the changes taking place in international environmental policy. These changes can be referred to in brief as 'policy innovation', 'multi-

P.P.J. Driessen and P. Glasbergen (eds.), Greening Society, 205–224.

actor' and 'multi-level governance'. The difference between international and domestic policy, and between the policy pursued by governments and private actors is constantly becoming less pronounced. Moreover, we observe that new forms of governance and a wider range of policy concepts and policy instruments are being used in the policies pursued on the environment internationally. These trends are illustrated in four cases: the climate negotiations, the approach taken regarding the pollution of the Rhine, the sustainable development treaties, and the realisation of the European Union's Fifth Environmental Action Programme. We then draw conclusions regarding the position and the role of the Netherlands in the multi-level field of environmental policy.

It should be pointed out here that these four cases are verifying cases. In other words: the Netherlands plays – at least on several aspects – an active role in each example. The objective of this casuistry is not so much to present a representative view of the role of the Netherlands in international environmental policy, but to give an impression of the role of the Netherlands as an innovator.

10.2 The Netherlands in an interdependent world

Environmental diplomacy is practised actively in the Netherlands. This, on the one hand, is due to internal factors – those in the Netherlands – and on the other, the 'autonomous' processes of Europeanisation and globalisation. In both this section and the next we argue that both types of factors – the internal and external factors – are starting to interact on an increasing level.

The direct Dutch environmental interest is a substantial argument for adopting an active international position. Our country's geographical location implies that a considerable part of the environmental load is of foreign origin (Bouwer, 1985). Pollution is carried into the Netherlands by air and water. World-wide problems, such as the depletion of the ozone layer and the intensified global warming effect, can have adverse consequences for the Netherlands. In other words, effective environmental measures at international level are essential if the Netherlands is to achieve its national, environmental goals. In addition to the immediate own interests there is also an awareness of one's own responsibility – or maybe even a Calvinistic awareness of wrongdoing – that plays a role in the way the Netherlands takes up its position with regard to the environment. The Netherlands not only imports pollution, it also exports it. The Dutch CO_2 emissions per capita are among the highest in the world. Also in the spatial sense is our ecological footprint a substantial one given that Dutch production and consumption goes hand in hand with a relatively heavy demand on space and the environment in foreign countries (Besselink, 1994). This awareness in Dutch society is fuelled by environmentalist and development organisations that hold a relatively strong position. Many Dutch people are members of these NGOs, and in turn, actively participate in the Dutch Government-subsidised consultation culture. In other words: the Netherlands is both the *victim* and *perpetrator* of transnational environmental problems. This double role is also referred to as 'ecological interdependence' (Glasbergen & Blowers, 1995; Hurrell & Kingsburry, 1992; Koehane & Nye, 1989; Mol & Liefferink, 1993).

Ecological interdependence is one reason for pursuing an international environmental policy, economic interdependence another. National economies are becoming increasingly interdependent and are being integrated in transnational markets. The European Union (EU) aims to achieve a single internal market within Europe and there is increasing talk in the world-wide dimension of a single world-wide market, something which is promoted actively by the World Trade Organisation (WTO). Processes of European integration ('Europeanisation') and 'globalisation' necessitate the mutual harmonisation of a large number of policy areas – including environmental policy – in order to guarantee freedom of trade as much as possible and to prevent the distortion of competition among nations. This applies a fortiori for a nation such as the Netherlands, with its open and strongly exports-oriented economy (Liefferink, 1997).

The processes of integration and globalisation not only have an economic, but also a political and cultural dimension (Waters, 1995; Yearley, 1996: Lubbers, 1998). The political dimension refers to the existence of (a progressively higher number of) international organisations and regulations, such as the EU and the UN. The fact that these international institutions exist means that there is also a stimulus to improve existing policy and formulate new policy. Such processes are also known as 'dynamic conservatism'. This dynamic conservatism is fuelled by the Netherlands given the nation's longstanding tradition as a proponent of consolidating international institutions, a tradition which relates to a variety of policy areas, certainly to environmental policy. For instance, in the National Environmental Policy Plan of 1989 it was argued that the United Nations Environmental Programme (UNEP) should be upgraded into a global environmental authority to possibly take even majority decisions (Ministry of Housing, Spatial Planning and the Environment, 1989, p. 181; Liefferink, 1997, p. 232).

The cultural dimension of integration and globalisation refers inter alia to the emergence of an international public opinion and transnational movements. Because these opinions and movements support certain (environmental) values they exert pressure on (inter) national politics to pursue more and better policy. Moreover, the enormous expansion we have seen in the ICT sector over the past few decades has intensified this trend since the Internet has increased the activating potential of NGOs (Kirn,

2000). By tradition, Dutch environmental NGOs have always played a major role in international umbrella organisations. The Dutch 'Stichting Natuur en Milieu' *(Foundation for Nature Conservation and Environmental Protection),* for instance, was the birthplace of the European Environmental Bureau (EEB) in the 1970s, the European umbrella organisation of environmental NGOs (Hey & Brendle, 1994). At the beginning of the new millennium this is still one of the driving forces behind the organisation.

The first conclusion we must draw from the foregoing is that the active role played by the Netherlands on the international environmental stage is not only prompted by the state-mediated 'general' interest in lowering the 'import' of pollution. The importance of an open and well-functioning system of international trade and competition, and thus a wide range of international actors, also plays a role. Additionally, the significance of the relatively strong Dutch environmental movement, both in national and international terms, should not be underestimated.

The second conclusion is that the actions undertaken by the Dutch Government in international environmental policy cannot only be explained by national factors. National political, economic and socio-cultural circumstances obviously form the basis for the actions of actors in the international arena, yet these actions only take shape in interaction with that same international arena. The argument can be defended that the strong economy-driven processes of Europeanisation and globalisation actually plays into the hands of an open economy such as the Netherlands. This can also partly explain the active international approach taken by the Netherlands, particularly in the field of the environment.

10.3 Policy innovation and multi-level governance

The connection between the national and international level of governance outlined in the previous section is, however, highly liable to change. In the traditional view, it is the community of states that jointly, in the absence of a coordinating authority, formulates environmental policy (Hocking & Smith, 1990). Juridical governance plays a major role in this respect, the treaty protocol approach being the main exponent (Benedick, 1991; Hurrell & Kingsbury, 1992; Rowlands, 1995). General objectives are formulated in so-called 'framework treaties' which are subsequently worked out into detailed measures in so-called 'protocols'. Brussels follows an approach similar to the treaty protocol approach, the so-called framework and daughter directives (Golub, 1998). In the traditional model it is the sovereign states that decide whether or not to enter into a treaty or protocol. The international decision-making process is thus generally geared towards reaching consensus and not majority decisions. As a rule this leads to vague compromises and decisions in which the slowest participant determines the group's pace, the so-called 'convoy principle'. Many international agreements also contain generic goals, means and time schedules in accordance with the formula of 'all nations must have reduced their emissions of X by Y% by the year Z in comparison with year P by implementing Q measures'. National governments must subsequently interpret these international obligations in their national legislation and regulations. However, this classic model of international environmental policy does no justice (or no longer) to the diversity among and within the different countries.

The classic model is thus subject to erosion. First and foremost, as a result of the constantly increasing ecological and economic interdependencies there is a change in the role of the sovereign states. This can be seen, for instance, from the fact that a supranational body such as the European Union is constantly gaining in status and political authority. The European Union has, or will have the power to take what will sometimes be radical decisions in many fields of policy. The greater part of the decisions taken on the environment have meanwhile been taken by majority vote, implying that individual Member States can be confronted against their wishes with international environmental measures (Liefferink, 1996). The status of the European Parliament has also been strengthened. The lead with regard to the economic aspects of globalisation has already been taken by an increasingly internationally operating trade and industry – multinationals and international interest groups. These parties are increasingly fighting for internationally harmonised, if not for more strict environmental standards (Kolk, 1999). Lastly, NGOs are becoming increasingly involved in international governance. Environmental NGOs are now also internationally oriented. Furthermore, they are also becoming more and more professional (Arts, 1998).

Another aspect of the innovation of the international policy arena is the modernisation of instruments (Golub, 1998). In addition to juridical governance, more emphasis has been given to economic and communicative governance in recent years. For instance: economic instruments such as levies and negotiable emission rights, and instruments based on dialogue and collaboration between the regulator and the parties to whom the regulations apply, such as agreements on future emission requirements or environmental reports drawn up by (large) businesses. As we explain in the following sections, EU environmental policy and international climate policy are important testing grounds for these new approaches. These new forms of governance give more attention to the aspects of differentiation, flexibility and openness. Policy can thus be better geared towards the possibilities in and the limitations of specific regions or industries. One interesting side effect of such 'customised policy' is that individual states start to play a less prominent role when implementing international agreements.

The outcome of these processes is that the difference between international and national policy gradually becomes more diffuse, and that the relative importance of national states starts to decline. 'International' environmental policy is becoming more of a process that takes place between a number of actors on different geographical and administrative levels. In short, traditional, hierarchic governance is now making room for multi-actor and multi-level governance (Kohler-Koch & Eising, 1999). Decisions are no longer automatically the domain of sovereign states alone. Nevertheless, this does not mean that the role of the national state has been played out completely, but policy is increasingly being effectuated in complex processes of collaboration between international institutions, regional and local authorities and – last but not least – private actors. This construction makes it more difficult to speak of 'the' position of the Netherlands in international environmental policy. On the other hand, it is quite possible to point out common tendencies in the actions taken by Dutch actors in the international policy arena. This is illustrated in the case studies set out in the following sections on the climate issue, the pollution of the Rhine, the new developmental relationship between the Netherlands and Costa Rica, and the interactive character of the Fifth Environmental Action Programme.

10.4 The Netherlands as 'innovator' in climate change policy

In the early 1980s, when human-induced climate change became framed as a policy problem, proposals to combat the problem were mainly based on the traditional regulatory approach, as usual in international environmental and, more particularly, atmospheric policy-making in those days (Bodansky, 1993). The dominant idea was that states - informed by experts - would jointly formulate general and legally-binding targets and timetables to reduce greenhouse gas emissions, to be included in an international treaty to combat global warming (Hurrell & Kingsbury, 1992). States should subsequently implement these at home. Although early policy responses reflect some of the above tendencies, such as the adoption of the stabilisation target for all industrialised countries in the Framework Convention on Climate Change (FCCC) in 1992, climate policy has evolved quite differently. In fact, the setting of targets and timetables has never become a top-down hierarchical road in the climate arena, with the global level as starting-point. The policy process has been multi-level in nature right from the start.

This becomes obvious if one views the international policy process from the perspective of a country such as the Netherlands. Although this country set its first domestic emission reduction targets in the late 1980s (3 to 5% reduction of carbon dioxide in 2000 compared to 1990 levels), new ones were determined by global agreements (FCCC, Kyoto Protocol) and European burden-sharing mechanisms ('EU bubble'). As a consequence, the Netherlands should reduce its emission of GHGs by 6% in the period 2008-2012 compared to 1990 levels, as a contribution to the common EU target of -8%, in order to adhere to the overall target of industrialised countries of -5,2%, as laid down in the Kyoto protocol (Gupta et al., 1998).

Climate policy, however, has not only had a multi-level character so far. Although some Asian and Middle-Eastern countries were reluctant to accept the input of non-governmental players in the climate arena, given their sometimes antagonistic relationships with NGOs at home, climate policy has also been multi-actor in nature right from the start. Due to the rather open participation processes of the UNCED, of which the rules of procedure were taken over by the FCCC, and the 'participatory style' of the international climate negotiations, non-governmental players, including many of Dutch origin, have always had access to the climate arena (Mintzer & Leonard, 1994). As a result, the role of epistemic communities, NGOs and business was significant (Arts, 1998; Boehmer-Christiansen, 1994; Kolk, 1999).

Besides having become more multi-level and multi-actor in nature, climate policy has also been innovated in terms of targets and instruments. First of all, targets have been differentiated even more (Gupta et al., 1998). Whereas in the FCCC (1992) there was only differentiation in target-setting between the developed countries (stabilisation target) and developing countries (no binding target), the Kyoto Protocol (1987) also provides for differentiation between the developed countries. Moreover, within the EU a region-specific differentiation of greenhouse gas reduction targets has been established (the so-called 'EU bubble'). For example, the Netherlands should reduce 6%, and the Germans 20%, whereas Greece may emit 20% more greenhouse gases in the period 2008-2012 compared to 1990 levels. Secondly, climate policy has become much more flexible (Jepma & Van der Gaast, 1999). Although industrialised countries are still bound to (differentiated) targets in the Kyoto Protocol, they are allowed more flexibility in achieving these, for example through voluntary agreements, covenants and flexible instruments.

The basic idea behind the flexible instruments is the following. Since the enhanced greenhouse effect is a global risk, it makes little difference - from a natural science definition of climate change - where emissions are reduced or removed. And if it does not matter where emissions are reduced or removed, one should do it, of course, where it is most effective and efficient. This may be the case for example in countries-in-transition or in developing countries, where the marginal costs of environmental investments are lower than in industrialised countries. This in turn implies that the responsibility of states and stakeholders for policy implementation, on the one hand, and the actual place where this is realised, on the other, are dissociated. Through *Joint Implementation* (JI) in Eastern Europe, the *Clean Development Mechanism* (CDM) in Third World countries, and emission trading among industrialised countries, western states and companies may reduce or remove emissions abroad. This offers them more flexibility in realising their goals, which - again - raises the chances of policy effectiveness.

The role of the Netherlands in the above innovations has been substantial. Particularly the following merits can be accredited to this country:

- The Netherlands was among the first countries to announce a domestic target on the reduction of greenhouse gas emissions at the end of the 1980s, a process which enhanced the setting of international targets (Mintzer & Leonard, 1994). In this respect, this country was an important 'initiator' of climate policy.
- The Netherlands introduced the concept of JI - although in terms of joint and cost-effective measures - already at the Noordwijk Conference on climate change in 1989 (Arts, 1998). These notions were mainly deduced from the Energy Charter of the late 1980s, and transferred to the climate talks. In this charter, West and Eastern Europe agreed to cooperate on making energy production and supply more efficient and, hence, more environmentally sound in the former communist states, among other things through western technology transfer and financial assistance. Later on, this idea evolved towards the concepts of JI as well as CDM in the Kyoto Protocol.
- The Netherlands was among those countries that promoted the concept of flexibility most strongly. Against the opposition of others (some industrialised countries, many developing countries, most NGOs), who feared a buying-off of real domestic change in terms of emission reductions, it fiercely defended a 50%-50% approach. At least half of the commitments should be achieved at home, according to the Netherlands, and the other half abroad through the flexible mechanisms.
- During its Presidency of the EU (1997) and of the Sixth Conference of the Parties to the FCCC (2000-01) respectively, the Netherlands played an important role in setting the 'EU-bubble' and in attempting to negotiate a compromise on elaboration of the Kyoto Protocol against the will of Bush administration.

Of course, one should not overestimate the innovative impact of a tiny country such as the Netherlands on multi-level and multi-actor negotiations. Other players were as important, or even more important, on certain topics. One might think of Norway and several East European and developing countries in establishing the mechanisms of JI and CDM. Moreover, the Netherlands also tried to slow down some innovations, such as the introduction of emission trading in climate policy by the USA. For long, the Dutch considered this kind of 'market environmentalism' opposite to their basic views on how to shape environmental policy-making.

Finally, one might question the motivation of the Dutch to promote certain innovations in climate policy. The overt motivation is, of course, to improve climate policy through differentiated targets and flexible instruments, thus increasing its legitimacy, efficiency and effectiveness. However, the covert motivation might be the prevention of painful, domestic change. As the Netherlands is one of those industrialised countries that failed to achieve the stabilisation target in 2000 (compared to 1990 levels) by far, let alone their initial domestic target of 3-5% reduction in the year 2000, it has become aware of the difficulty to attain emission reductions at home (Ministry of Housing, Spatial Planning and the Environment, 1999). This implementation deficit is probably another reason for the Dutch to have propagated the flexible instruments and the 50%-50% approach so fiercely.

10.5 Flexibility and the increasing level of openness in the Rhine negotiations

Since 1950, the Rhine riparian states Switzerland, France, Germany and the Netherlands together with Luxembourg, located on the Moselle, have been engaged in negotiations on how to tackle the problem of pollution (Dieperink, 1997, 2000). The object of these negotiations in the International Commission for the Protection of the Rhine (ICPR) were the discharge of chloride by the French potassium mines and the discharges from the chemical industry (in the Ruhr area, the Neckar area and in the vicinity of Basel). Both issues were reason for the Netherlands in particular, located downstream, to enter into a long-term intergovernmental process of negotiation, in which there was often no scope to negotiate because of the inflexible attitude taken by the riparian states. The negotiations on the salt discharges from the French potassium mines had for more than 20 years resulted in a series of moves in which the theoretical possibilities of reducing the impact of these discharges (transport by ship or through pipelines into the sea, injection into the soil as an industrial process) were simply repeated on a regular basis without consensus being achieved. The

Netherlands adopted a position on principle, demanding that France achieved a permanent reduction in the discharge in question. The Netherlands also took the stand that the discharge of chemical substances should be brought to an end in the very near future too. The negotiations on chemical pollution came to a close in the 'Convention for the Protection of the Rhine Against Chemical Pollution', a framework treaty that should have been worked out in several more detailed industry-specific and content-specific protocols, which would then form the basis for the discharge permits to be issued in the catchment area. The fact that the European Union played a pivotal role in elaborating these regulations was not particularly beneficial in terms of speed.

The Rhine riparian states changed their attitude with regard to the salt discharge and chemical pollution in the mid-1980s, the Netherlands playing a key role in both cases. The expensive proposals put forward by France for a permanent reduction of these discharges, to be financed by all the riparian states jointly, was rejected by the Netherlands. Instead, the Netherlands put forward a cheaper proposition in which the wishes expressed by the French to store the salt on a temporary basis and to discharge it in a controlled fashion at high tide were respected. This was set out in the 'Rhine Chloride Protocol'. The second part of the package agreed by the other Rhine riparian states was to prevent a collection point for drinking water in the IJsselmeer from becoming brackish.

The approach to the chemical pollution was brought into a total package, the 'Rhine Action Programme' (1987). By implementing this programme, the Rhine riparian states not only attempt to bring the (drinking) water quality up to an acceptable level, but also make an effort to recuperate the ecology of the Rhine catchment area. The return of salmon to the Rhine, the envisaged target of this programme, has been worked out in objectives relating to the pollution load reduction and in objectives in the field of hydraulics and morphology of the catchment area (i.e. spawning locations and fish ladders). The Rhine riparian states now have greater freedom to specify measures for improving water quality,. Only the objectives themselves have been harmonised (in the form of aims to reduce the pollution load). The riparian states themselves may choose an adequate set of measures.

Running parallel with the increasing amount of flexibility, the dialogue on the Rhine pollution became more open in the 1990s. The NGOs played no role at the actual negotiating table in the 1970s and 1980s. Environmental organisations and drinking water companies placed emphasis on activities geared towards the public and on instituting legal proceedings. The industry landed up in the dock and was not very open about the nature of its discharges. Today, NGOs play an active role in innovating the policies of the International Commission for the Protection of the Rhine (ICPR). They have the status of observer in the plenary ICPR meeting, showing evidence of a higher level of openness. By convening conferences and popularising the ecological recuperation, the ICPR contributes towards increasing this level of openness.

The Dutch Government has adopted a highly active role in the negotiations. In order to accelerate the negotiations on the salt discharges, the Dutch Government organised ministerial conferences and even started to make a financial contribution towards the French research efforts and clean-up operations. When the conflict with France escalated, the Dutch Government took the highly unusual step of calling back its ambassador to Paris. The Dutch Government also reacted alertly to the disasters that occurred upstream. The explosion at the Swiss chemicals giant, Sandoz, in 1986 resulted in massive fish mortality. The Dutch Government made use of the public commotion to call for a Rhine Action Programme. The paradigm shift under way at the time in policy pursued at national level (flexibility, differentiation, openness) was resounded resolutely in the draft text of the Rhine Action Programme drawn up by the Dutch Government. Moreover, the Dutch Government was investing a great deal in monitoring quality and was thus well able to underpin its ideas.

The efforts of the joint Rhine riparian states were not without result. The Rhine has been changed from what in biological terms was a virtually 'dead' river into one in which the pollution has been reduced considerably, where salmon can thrive once again. Despite the fact that the Netherlands fulfilled a key role in the negotiating process as the 'initiator', it was certainly not the most progressive riparian state where the actual clean-up operation was concerned. Measures were more likely to be taken earlier upstream than in the Netherlands.

10.6 Innovation in policy on the environment and development aid

There is a long-standing imbalance in the relationship between western donor nations and recipient developing countries. This is mainly reflected in the one-way traffic in the financial and normative sense. Developing countries experience this as paternalism. Developing countries are also critical of the predominant western perspective of many international environmental treaties. These countries have called for a more balanced cooperation for some time now, as also have international NGOs. At the beginning of the 1990s we see the criticism leading to a paradigm discussion. Thinking in terms of top-down aid to the developing countries slowly changes to thinking in terms of cooperative regimes. The 'ownership issue' of development aid programmes, the question of who is actually responsible for structural changes, starts to be answered in a different way. Development aid starts to be seen less as a primarily technical and economic matter that must be achieved through external, strongly hierarchically conditioned aid, but more as support for a transformation process initiated by communities as a whole. An analysis conducted by Hein (2000) shows us that the World Bank especially (and to a lesser extent the IMF) adopted this paradigm as its main focus of attention in the second half of the 1990s.

An early and far-reaching operationalisation of this paradigm is the subject of a sustainability treaty developed by the Netherlands during the work carried out in preparation of the UNCED Conference (1992) (Glasbergen & Miranda, 2000). A form of long-term environmental development aid is envisaged which – while it will be concluded by governments – will be detailed by individual communities on the basis of equality and reciprocity. Collaboration will then need to be shaped on the basis of participation of NGOs in all the countries concerned. To work out the actual details of this idea, the Netherlands sought for partners who, in terms of population figures and/or territory, are comparable with the Netherlands, are politically stable, are undergoing democratic development, and show a clear commitment towards sustainable development and nature conservation. The programme as a whole should be experimental and could serve as a model for other countries.

With a view to achieving a certain spread of experiences, bilateral, Sustainable Development Treaties (SDTs) have been concluded with Costa Rica, Benin and Bhutan. The declarations of intent were signed at the UNCED Conference and were given an enormous amount of publicity. These SDTs were also promoted at a later date by the Netherlands at international conferences and workshops. The type of treaty we are talking about here is both innovative and challenging, focusing not only on the complex issue of sustainable development, but also placing this aim into what up to then had been an unprecedented open context. The SDT could be a supplement, and sometimes even an alternative for existing environmental development treaties. Although no other nation has followed the initiative to date, there is still enough interest being shown by the international community.

The SDT between the Netherlands and Costa Rica, which we briefly explain below, attempts to give meaning to the concept of sustainable development. It sets out a framework for detailing the concept of 'world-wide partnership' that was central at the UNCED Conference. The essence being the notion of a shared interest in sustainable development by both the developed and the developing countries – each having their own, but connected specific views. The build-up of this cooperation is based on equality, reciprocity and participation. The treaty is not exactly specific in setting out how the goals are to be achieved, sufficing with a summary of sixteen non-equivalent topics which are not prioritised by either party. Reference is made inter alia to promoting consumer patterns that protect the ecological basis of development, the sustainable management of natural resources, the control of hazardous waste streams, a reduction in CO_2 emissions and energy saving, and attention given to the environmental effects of international trade relations.

Whereas there is tangible evidence of a relationship of dependence (the developing country is completely dependent on the donor in the financial sense), the link made between sustainable development, reciprocity, equality and participation, places the treaty in a different political and ideological context. This context will need to be given operational substance; the most important task for the sustainability treaty. Since the beginning of 1999 a total of 62 new projects have been started up in Costa Rica. The Costa Rican NGO umbrella organisation 'Fundecooperation' has assumed the role of coordinating body and its activities are subsequently inspected by the Dutch Embassy. These projects are regarded as crucial in the aim towards sustainable development. Major initiatives have been taken in terms of promoting agro-biodiversity; improving agricultural production methods (to ensure that they meet the Max Havelaar standard); the involvement of the local population in eco-tourist projects and the management of nature reserves; the introduction of environmental technology in small-scale production processes; recycling; and improving energy efficiency. The operational significance attached to the concept of sustainable development is reflected in these projects.

However, the treaty's implementation is problematic in the Netherlands. Ecooperation, the Dutch NGO umbrella organisation, has done very little in the way of informing the Dutch population of this joint venture (Glasbergen & Miranda, 2000). People in Costa Rica feel that there is too little interest shown in the Netherlands as to what is being achieved in their country. Attempts to establish contact with the Dutch NGOs by the Costa Rican NGOs have been unsuccessful. Ecoorporation was evidently incapable of attracting the interest of NGOs in the treaty. The conclusion drawn in Costa Rica is that the Dutch sectors (NGOs, universities, trade and industry, the local authorities) stick to their own sector and have not yet – as is now the case in Costa Rica – learnt to join forces and pull together.

During the period of the treaty's implementation we observe an increasing level of activity around issues of sustainable development in Costa Rica and a reduction in the activities in the Netherlands. In Costa Rica there is a learning process under way in which capacities are being built up. This is not the case in the Netherlands, or to a much lesser extent. Also, apart from a few (chiefly) study projects, there is a lack of bilateral projects that should be started up by the Netherlands. The principle of reciprocity has therefore not been operationalised.

Furthermore, political support for the treaty is also diminishing in the Netherlands. The present Minister for Development Cooperation, Herfkens, even wishes to terminate the treaty. And so we see that what is basically an innovative international environmental instrument has been sidetracked, at least in the Netherlands.

10.7 Export of the interactive approach to Brussels: the EU's fifth environmental action programme

Since the early 1970s, the EU has published Environmental Action Programmes. Although not strictly binding, these Programmes have been important in guiding policy development in Brussels. Starting as hardly more than broad 'shopping lists', the subsequent Action Programmes evolved into documents with considerable strategic ambition. This development culminated in the Fifth Action Programme, covering the period 1993-2000 (European Commission, 1993). In its design, the Netherlands played an important role.

In the run-up to the 1992 Rio Conference, political attention at the international level for environmental issues was particularly high. At the same time, the European Commission, and more specifically its Directorate-General for the Environment (DG XI), became increasingly dissatisfied with the results of the earlier Action Programmes. This made the Commission look for ways to formulate the next issue in the series as a more coherent and comprehensive framework for future policies (Kronsell, 1997).

In 1989 the Dutch National Environmental Policy Plan (NEPP) had been published (Ministry of Housing, Spatial Planning and the Environment, 1989). It was internationally acclaimed as the most comprehensive plan for environmental management up to then. The Dutch government was eager to export its 'show-piece' to other countries and, if possible, to international organisations (Liefferink, 1999).

Considering in addition the Commission's general tendency to look for examples for its policies at the Member State level (Héritier, 1996), the ambitions of DG XI and the Dutch government almost naturally came together in a strong role of the Dutch in the preparation of the Fifth Action Programme, most conspicuously in the form of the appointment of an experienced Dutch civil servant as one of the principal authors. This process was further facilitated by the fact that DG XI at the time was also led by a Dutch, Laurens-Jan Brinkhorst.

The Dutch influence on the Programme is immediately evident in its structure. Similar to the NEPP, chapters are built around themes and target sectors. More fundamentally, the Programme is pervaded by an interactive approach to environmental policy-making. One of its key notions is 'shared responsibility'. This not only refers to striking the optimal balance between the European, the national and the sub-national level, reminiscent of the principle of subsidiarity, but also to a close interplay between public and private actors. The latter aspect is worked out, among other things, by stimulating a dialogue with industry and other target sectors and encouraging voluntary agreements and self-regulation.

To what extent has the approach advocated by the Fifth Action Programme been successful? Fully fledged forms of 'joint environmental policy-making', such as voluntary agreements, have hardly developed at the EU level. This can be explained with reference to various legal complications as well as the basically non-corporatist character of public-private relations in Brussels (Mol et al., 2000). Also in a more general sense, however, the interactive approach of the Fifth Action Programme was criticised. The official assessment of the Programme, issued in 1999, states that 'practical progress towards sustainable development has been rather limited, mainly because there was no clear recognition of commitment from Member States and stakeholders...' (European Commission, 2000, p. 9). Not surprisingly, the draft of the Sixth Environmental Action Programme returns to a somewhat more traditional view on policy making. A key priority is the implementation of existing legislation. The draft still stresses the importance of 'working with the market', but the emphasis has moved back from dialogue and shared responsibility to a wide range of carrots and sticks (European Commission, 2001).

The transplantation of the NEPP's essential way of thinking to the EU level appears to have been in many senses 'a bridge too far'. The case shows that the export of basic policy concepts only makes sense if differences in the political and institutional context and the prevailing policy culture are duly taken into account.

10.8 The four cases compared and analysed

From the four cases it becomes obvious that the Netherlands has succeeded in playing the role of 'innovator' in at least some domains of international environmental policy. Its main merit in climate policy was the introduction of the idea of joint implementation (1989); in the river Rhine policy the introduction of a flexible and participatory approach in the Action

Programme (1987); in development co-operation the introduction of the bilateral Sustainable Development Treaty (1992); and in EU environmental policy the introduction of the interactive approach in the Fifth Environmental Action programme (1995). This seems to legitimise the conclusion that the Netherlands is indeed a pioneer in international environmental policy (Andersen & Liefferink, 1997). At the same time, there are a number of reasons to put this conclusion into perspective (besides the fact that it is very difficult to single out the role of one country in multi-actor and multi-level policy processes). These reasons are related to:

- motivation(s) of the Netherlands to achieve the status of pioneer;
- attainment and effectiveness of its innovative proposals;
- implementation deficit of environmental policy in the Netherlands;
- the maintenance of unsustainable production and consumption patterns in this country; the growing reluctance of the Netherlands to maintain its forerunner position in the near future.

One may ask why the Netherlands is so keen on 'exporting' environmental policy innovations. One reason is, as both the climate and EU case show, to improve policy performance. By introducing the interactive approach in the EU, the Dutch hoped to increase the effectiveness of European environmental policy. And through joint implementation, they hoped to increase the efficiency of climate policy, thus making it more attractive to laggards in the climate negotiations. At the same time, the latter case also shows that the Netherlands' focus on flexible mechanisms was inspired by preventing painful economic measures at home. Hence, motivations for being an 'innovator' are not always that benign. The Rhine case, next, shows that the Dutch are also driven by the intent to solve their domestic environmental problems through international policy innovation. By introducing a new, more flexible package deal they hoped to persuade the French to reduce the draining of salt into the Rhine, something that failed in earlier, more 'traditional' policy initiatives. Another motivation to be an 'innovator' is international exposure, which is so important for a small country that wants to be taken seriously. That motivation was probably one of the main backgrounds of the introduction of the Sustainable Development Treaty. With much noise, the Dutch announced its treaties with Bhutan, Benin and Costa Rica at the UNCED, as if it would say: 'Look here, we, the Dutch, from a very small country, are fundamentally rewriting the traditional approach of development co-operation from a sustainable development perspective'. All in all, one has to conclude that the role of innovator is not only chosen for the 'common good' (to improve policy performance), but for 'egoistic motivations' as well (preventing painful measures, national problem solving, international exposure). 'Pioneering' then coincides with 'furthering national interests'. True, that is not necessarily wrong from a normative perspective, but definitely something that should be taken into consideration.

A second issue that adds nuance to the pioneering role of the Netherlands is the lack of attainment and, in the end, of effectiveness of its innovative proposals. The best - or better: worst - case is the Sustainable Development Treaty. Although to some extent successful in Costa Rica, it seems to have become a dead end in Dutch development co-operation. Probably, this type of treaty is too much wishful thinking and ideological to be attainable and effective in the end. After all, a donor-recipient relationship will always remain one of one-sided dependency, irrespective of any treaty based on the principle of equity. A similar lack of attainment and effectiveness is related to the interactive approach in EU environmental policy. The Dutch 'Poldermodel' (see chapter 9) seems not really to be working at the European level. More successful is the flexible approach the Dutch have propagated in Rhine and climate policies, although the concept of joint implementation is only a success on paper so far, and still needs to prove its effectiveness on the ground (and there are many doubts about that). Given this mixed picture, it seems that the Netherlands is foremost a 'discursive innovator', very good at introducing new concepts, but definitely less a true innovator of policy practices. This may at the end of the day frustrate the pioneering role, and the underlying motivations, the Dutch have striven for (at least in the past, see below).

A third issue, that might also put the pioneering role of the Netherlands into perspective is a certain reputation of the Dutch of being hypocrite: 'saying to others what they should do on the one hand, while falling short in implementing domestic and international policies on the other'. Examples of Dutch implementation deficits are: the EU Habitat Directive, the EU Nitrate Directive and the Climate Convention (FCCC). With regard to these three, the Dutch failed in implementing its main domestic commitments so far. Therefore it is no surprise that the Netherlands was summoned by the European Court of Justice several times in the past. Yet this blame to the Netherlands is not entirely fair, because compared to other countries the Dutch are doing quite well as far as the implementation of environmental policy is concerned. Moreover, the implementation deficit is a general and world-wide phenomenon in international environmental policy.
Related to this third issue is a fourth one: the maintenance of unsustainable production and consumption patterns in the Netherlands, such as intensive, large-scale agriculture and huge volumes of (air) traffic. Critics might argue that the Dutch have no right to speak up in international environmental policy as long as they maintain unsustainable development patterns at home. Although the critics are probably right that this should inspire the Netherlands to take a less pretentious attitude in the

international arena, they are wrong insofar as they overlook the fact that this is a general problem, valid for other countries as well. Moreover, the transition to sustainable development is a long-term process that cannot be achieved overnight, and should not force countries to be silent about policy innovations in the short run.

A final issue that put the pioneering role of the Dutch into perspective is its own reluctance to play this role so explicitly in the near future. Inspired by the above experiences and criticisms, the Netherlands has become more careful in what to propose and in what not to propose in international forums. Therefore one might argue that the pioneering role of the Dutch in international environmental policy is declining.

10.9 Conclusion

Policy innovation in the Netherlands and in the international dimension are partly linked. In addition to international treaties and European guidelines with generic measures and time schedules and classic instruments of governance, the community of states – in the negotiations with private actors – are increasingly changing over to introducing different goals, flexible instruments, as well as communicative and economic governance. The casuistry in this chapter illustrates that in a number of these policy innovations the Netherlands has played a (modest) innovative role. The Netherlands has been a successful 'international innovator' in several cases, particularly in terms of concept.

The foregoing legitimates the conclusion that the Netherlands is still a pioneer in international environmental policy. Yet at the same time we must realise that this role is under pressure. The reasons for this are: (1) the occasional motives of pure self-interest for the Netherlands to fulfil this role; (2) the lack of feasibility and effectiveness of many of the Dutch innovative proposals; (3) a lack of implementation and a chiefly non-sustainable development in the Netherlands itself; as also (4) the current reticence of this nation to be an international innovator any longer.

References

Andersen, M.S. & D. Liefferink (eds.) (1997) *European Environmental Policy. The Pioneers,* Manchester University Press, Manchester.

Arts, B. (1998) *The Political Influence of Global NGOs, Case Studies on the Climate and Biodiversity Conventions,* International Books, Utrecht.

Bennedick, R.E. (1991) *Ozone Diplomacy: New Directions in Safeguarding the Planet,* Harvard University Press, Cambridge.

Besselink, C. (1994) *The Netherlands and the World Ecology,* The Netherlands Committee for IUCN, Amsterdam.
Bodansky, D. (1993) 'The United Nations Framework Convention on Climate Change: A Commentary', Yale Journal of International Law, Vol. 18, No. 451, pp. 453-559.
Boehmer-Christiansen, S. (1994) 'Scientific Uncertainty and Power Politics: The Framework Convention on Climate Change and the Role of Scientific Advice', in: B. Spector, G. Sjöstedt & I. Zartman, *Negotiating International Regimes. Lessons Learned from the United Nations Conference on Environment and Development (UNCED),* Graham & Trotham/Martinus Nijhoff, London, pp. 171-180.
Bouwer, K. (1985) 'Grensoverschrijdende milieuproblemen in een geografisch kader', *Geografisch Tijdschrift,* 19 (15), pp. 385-398.
Dieperink, C. (1997) *Tussen zout en zalm, Lessen uit de ontwikkeling van het regime inzake de Rijnvervuiling,* Thesis Publishers, Amsterdam.
Dieperink, C. (2000) 'Successful International Cooperation in the Rhine Catchment Area', *Water International,* vol. 25, no 3, pp. 347-355.
European Commission (1993) 'Towards Sustainability: Environmental Action Programme 1993-2000', *Official Journal of the European Communities,* C 138, May 17.
European Commission (2000) *Global Assessment. Europe's Environment: What Directions for the Future?,* Office for Official Publications of the European Communities, Luxembourg.
European Commission (2001) *Environment 2010: Our Future, Our Choice. Sixth Environmental Action Programme (draft),* European Commission, COM(2001)31, Brussels.
Glasbergen, P. & A. Blowers (eds.) (1995) *Environmental Policy in an International Context. Part I: Perspectives,* Arnold, London.
Glasbergen, P. & M. Miranda (2000) 'Het duurzaamheidsverdrag tussen Nederland en Costa Rica; een nieuwe oriëntatie op milieu- en ontwikkelingssamenwerking', *Milieu. Tijdschrift voor Milieukunde,* vol. 15, no. 5, pp. 226-240.
Golub, J. (1998) *New Instruments for Environmental Policy in the EU,* Routledge, London.
Gupta, J., C. Jepma & K. Blok (1998) 'International Climate Change Policy: Coping with Differentiation, *Milieu, Tijdschrift voor Milieukunde,* vol.13, no. 5, pp. 264-274.
Hein, W. (2000) *Global Governance and the Evolution of a New Role of International Financial and Economic Institutions for Sub-Saharan Africa,* draft.
Héritier, A. (1996) 'The accommodation of diversity in European policy-making: regulatory policy as a patchwork', *Journal of European Public Policy,* vol. 3, no. 2, pp. 149-167.
Hey, C. & U. Brendle (1994) *Umweltverbände und EG. Strategien, Politische Kulturen und Organisationsformen,* Westdeutscher Verlag, Opladen.
Hocking, B. & M. Smith (1990) *World Politics. An Introduction to International Relations,* Harvester/Wheatsheaf, New York.
Hurrell, A. & B. Kingsbury (1992) *The International Politics of the Environment,* Oxford University Press, Oxford.

Jepma, C.J & W. van der Gaast (eds.) (1999) *On the Compatibility of Flexible Instruments,* Kluwer Academic Publishers, Dordrecht.

Keohane, R.O. & J.S. Nye (1989) *Power and Interdependence. World Politics in Transition,* 2nd ed. Scott, Foresman & Company, Glenview.

Kirn, W. (2000) 'Globalization: The New Radicals', *Time,* April 24, 2000, pp. 34-38.

Kohler-Koch, B. & R. Eising (eds.) (1999) *The Transformation of Governance in the European Union,* Routledge, London.

Kolk, A. (1999) 'Multinationale ondernemingen en internationaal klimaatbeleid'. *Milieu. Tijdschrift voor Milieukunde,* vol. 14, no. 4, pp. 181-91.

Kronsell, A. (1997) 'Policy Innovation in the Garbage Can: the EU's Fifth Environmental Action Programme', in: D. Liefferink & M.S. Andersen (eds.) *The Innovation of EU Environmental Policy,* Scandinavian University Press, Oslo, pp.111-132.

Liefferink, D. (1996) *Environment and the Nation State. The Netherlands, the European Union and Acid Rain,* Manchester University Press, Manchester.

Liefferink, D. (1997) 'The Netherlands: A Net Exporter of Environmental Policy Concepts', in: Andersen, M.S. & D. Liefferink (eds.) *European Environmental Policy. The Pioneers,* Manchester University Press, Manchester.

Liefferink, D. (1999) 'The Dutch National Plan for Sustainable Society', in: N.J. Vig & R.S. Axelrod (eds.) *The Global Environment, Institutions, Law and Policy,* Earthscan Publications Ltd, London.

Lubbers, R.F.M. (1998) *Governance in an Era of Globalization,* October 16, http://www.globalize.org/govern1.html

Ministry of Housing, Spatial Planning and the Environment (Ministerie van VROM) (1989) *National Environmental Policy Plan: To Choose or to Loose,* Den Haag.

Ministry of Housing, Spatial Planning and the Environment (Ministerie van VROM) (1999) *International Peer Review of Dutch Environmental Policy,* Publicatiereeks Milieustrategie 1999/17, Den Haag.

Mintzer, I.M. & J.A. Leonard (eds) (1994) *Negotiating Climate Change. The Inside Story of the Rio Convention,* Cambridge University Press, Cambridge.

Mol, A.P.J. & J.D. Liefferink (1993) 'European Environmental Policy and Global Interdependence: A Review of Theoretical Approaches', in: Liefferink, J.D., P.D. Lowe & A.P.J. Mol (eds) *European Integration and Environmental Policy,* Belhaven Press, London, 1993.

Mol, A.P.J., D. Liefferink & V. Lauber (2000) 'Epilogue: Conclusions and Policy Implications', in: A.P.J. Mol, V. Lauber & D. Liefferink (eds) *The Voluntary Approach to Environmental Policy. Joint Environmental Policy-Making in Europe,* Oxford University Press, Oxford, pp. 217-226.

Rowlands, I.H. (1995) *The Politics of Global Atmospheric Change,* Manchester University Press, Manchester, 1995.

Waters, M. (1995) *Globalization,* Routledge, London.

Yearley, S. (1996) *Sociology, Environmentalism, Globalization,* Sage Publications, London.

PART 5

CONCLUSIONS

Chapter 11

THE PARADIGM SHIFT IN ENVIRONMENTAL POLICY RESEARCH

Pieter Leroy and Nico J.M. Nelissen

11.1 Policy studies and policy in practice

Scientific research is seldom carried out for the sole purpose of describing reality. The intention is almost always to use scientific insight in an attempt to forecast the actual situation and, if at all possible, to change it. This applies in particular to policy studies, a variety of social science disciplines that is difficult to define, but that embraces disciplines such as sociology, political science, public administration, economics and others that have 'policy' as the subject of their research. These disciplines not only focus their research *onto policy*, but they also conduct that research *for policy*. After all, policy studies – more than the social sciences in general – are characterised specifically by the objective of intervening in societal reality.

Conversely, society, politics and policy formulate a variety of the needs for knowledge as a result of the social and policy problems they are confronted with: knowledge about the nature and causes of problems, and especially knowledge on how they should be tackled. In addition to the theoretical considerations, these practical needs for knowledge are among the main points of departure for scientific research both *onto* and *for* policy.

Accordingly, policy studies and policy practice constantly interact. While each has its own institutional pattern, its own rules and structures, both have a mutual influence on the other. Policy's influence on science becomes structurally manifest in science policy, yet it can also be strategic through the funding of research programmes and projects. Policy studies can also advise on the main lines of the content, the strategy and the organisation of a field of policy or the constituent parts thereof. In general,

P.P.J. Driessen and P. Glasbergen (eds.), Greening Society, 227–244.

however, the element of reciprocal influence is more subtle and ensues from the fact that, despite the 'separate' circuits of science and policy, the agendas of policy studies and policy practice share a number of the same themes.

These considerations also apply to those policy scientists engaged in work on environmental issues. These scientists do not generally wish to limit themselves to the analysis and clarification of their subject, but also wish to make a contribution towards the innovation and improvement of environmental policy. By contrast, their agenda, their prioritised issues and preferential themes are also co-determined by the policy questions of everyday life. Thus environmental policy research also bears the marks of the researchers' ambitions to 'steer', the constant interaction with social, political and policy developments, and the ensuing needs for knowledge.

In this chapter we take a look at the questions that have been dealt with to date in environmental research carried out within the scope of policy studies, and we look at which perspectives were the most often used and which prioritised themes were the ones most often studied. We obviously give attention to shifts in the environmental research concerned. It will become evident that these shifts in science activity reflect to a large extent the paradigm shift that has taken place in environmental policy in practice.

We first give a short account of the context within which environmental policy research should be seen before moving on to explain the research itself on the basis of three distinguishable and more or less successive dominant perspectives and themes that were used.

11.2 Shifts in the context of environmental research

As already pointed out, environmental policy research is not conducted in a vacuum but in close interaction with society and politics, and with policy in general and environmental policy in particular. Therefore, we investigate in this section a few of the changes that have taken place within the context of environmental policy research. We look in particular at three separate aspects of that context, especially at the changes that have taken place over the past few decades, devoting attention to:

- how environmental problems are formulated in the societal and the political context, how they are placed on the agenda and how they are constructed;
- the views on policy and management and how the two are practiced, particularly in terms of the role played by government and other societal actors;
- the gradual shifts in views within policy research in general, and on aspects of management in particular.

The societal and political formulation of environmental problems and how they are placed on the agenda

Virtually all Western nations placed environmental problems on their societal and political agendas in the early 1970s. This went hand in hand with a rapid growth in environmental awareness; the environmentalist movement emerging on the wave of that awareness, and environmental policy developing as an autonomous field of government policy. Nevertheless, there has still been a considerable shift in the way society and politics have perceived and formulated environmental problems over the past thirty years. Initially, environmental problems were defined chiefly as technical environmental pollution problems that could be fought by government-imposed technologies for soil-cleaning, filtering, combining or diluting operations. Scientific insights, especially standards based on health and occupational risks, formed the basis for a policy that consisted mainly of rules relating to the maximum amount of emission or immission of hazardous or polluting substances. The promulgation of those rules, their operationalisation in a variety of permit systems and so forth were virtually the tasks of government alone.

It gradually became clear in the 1980s, however, that this policy strategy was less successful than had been envisaged. This was partly because of the poor implementation of environmental policy (we return to this later), and partly due to shifts in how the problems were perceived. The environmental problems were apparently more serious in terms of cohesion, scale and complexity than had been thought originally: it was impossible to bring them down to the local problems of air, water or soil pollution, nor could they be reversed simply by technological means, let alone by end-of-pipe technology. Placing the 'acidification' issue on the agenda, for instance, instead of the till then dominating 'air pollution' issue, implied the recognition of complex associations of different sources, different types of pollution and diverse effects. By constructing such environmental problems as 'acidification', and particularly 'climate change', and by placing these items on the agenda, the continental and later on even the global character became apparent. Dinkelman (1995) illustrates how physically highly similar circumstances were referred to as 'air pollution' in the 1970s, 'acidification' in the 1980s and in the 1990s as 'climate change'. Obviously we are not concerned here with the labelling in itself, but with the fact that the shifts in perception and formulation refer, inter alia, to cause and consequence mechanisms, to other involved actors, and to other levels of scale, both geographically and politically. The shifting societal and political constructions of environmental issues repeatedly implied new propositions for policy development in terms of content, strategy and process, thus leading to a constant demand for new knowledge and to new questions being put to the scientific community.

Shifting views on the responsibility of government and other actors
Environmental policy emerged in the political context of the early 1970s. In that period there was an enormous belief in the makeability of society and in government's managerial capacities. Environmental policy was seen as one of the new fields of policy that the welfare state would develop further. The emerging environmental policy was biased by this political context: a predominant choice for a state-initiated policy, implemented on the one hand through collective services (water purification, waste disposal and other services), and by way of command and control strategies laid down in legislation and regulations on the other. The steering capacities of the government on the one hand and the relative compliance of market and society agencies, of the business community and citizens on the other, were considered no more than logical.

On grounds of either facts or normative and ideological motives, belief in the government's managerial capacity declined substantially in the 1980s. The government should not, and indeed must not, have so many managerial ambitions, and societal and market actors would have to assume more responsibility. The policy-making process, including that for environmental policy, should be less of a top-down process and should involve a higher level of consultation. In consequence, it should be orchestrated less by way of prohibitions and more by economic and communicational means. In this sense, the government had more of a supporting and facilitating than a directing role. Certain aspects led to heated debates on the role of government and 'the management of society' in general, debates that unravelled under themes such as 'from government to governance'. Yet it was not only the views that changed, these debates also had demonstrable practical consequences and led to what were sometimes spectacular operations involving the privatisation and liberation of what at that time were government tasks, e.g. in the field of communications, transport and the infrastructure. They also led to a much more cooperative style of policy application in which command and control had to make more room for inter-organisational management via networks and more scope for consultation with stakeholders aiming to achieve consensus.

Trends in policy studies
Policy research in the environmental field is, quite naturally, also influenced by the trends in policy studies in general. In turn, the latter are influenced by the changes in views and practices described in the foregoing regarding the role and responsibility of government and other societal actors. Policy studies are in more than one sense the result of the welfare state and the criticism of that state. After all, the discipline of policy studies has developed so successfully partly because it propagates the good news of the welfare state: 'the good society' could be brought closer if the

government acted wisely. This was the dominant, rationalistic view at the time. In that context, policy researchers felt it their duty to deliver sound problem analyses, to work out policy strategies, to adjust governmental organisation, and to supervise the process. Policy studies thus supported the ambitions of the welfare state to intervene.

Certainly with regard to the government's role, those ambitions have since then substantially weakened. And yet many of today's policy researchers are professionals who are not free of obligation but are committed to policy. In other words, they are not only interested in the analysis, but also in the design and application of new management strategies and new instruments of intervention adjusted to the new political concept and the new styles of management. The two sides of policy studies (the science *of* policy and science *for* policy) are even today constantly perceptible.

11.3 The shift in the approaches used in environmental policy research

In the foregoing we have painted a very general picture of the context in which environmental policy studies research has developed over the past few decades. There are two reasons why this picture is far too general. Firstly because, apart from the three factors set out in the above, there are numerous other factors that also influence environmental policy research, such as the institutional position and strength of either policy sciences or environmental studies at universities. Secondly and consequently, because both the context and the trends in environmental policy research differ from country to country. The societal and political attention for environmental issues, their institutional context, the interaction between policy practices and policy studies, the state of the art of the latter, et cetera, differ to a very large extent throughout Europe. This, as we see from some comparative research on environmental policies (cf. Liefferink, 1996; Haverland, 1998; Weale et al., 2000), leads to widely different practices in policy that are paralleled by substantial differences in environmental policy sciences' activities.

The following picture is based very much on the Dutch situation. It thus relates to a country with a relatively strong social and political environmental awareness, a traditionally strong belief in collective responsibility and governmental management, and a substantially developed and reasonably pragmatic social science tradition. A large part of Dutch environmental policy research is therefore the direct result of specific questions asked by government and government assignments. It is therefore, as we will conclude later on, *research for policy* that predominates in environmental policy studies, probably even more than in general policy studies. The influence of that research on policy is also demonstrable (cf. Smits & Ringeling, 1997). The context of a strong interaction between environmental policy and environmental policy research is typical of the Netherlands. Whereas the situation is certainly different in the UK, France or Italy, for instance, and even in the neighbouring countries of Germany and Belgium, we will point out any comparable developments in other countries where relevant.

Developments in Dutch environmental policy research can be arranged schematically in three more or less consecutive dominant methods of approach (or paradigms). These dominant approaches – and this is not by accident – run reasonably parallel with the shifting formulation of environmental issues and the changing attitudes regarding the responsibilities of government and other actors. Although the observed approaches in actual fact merge and overlap, they still show some distinct differences with regard to the underlying views on environmental issues, to the most opportune strategy to be used for tackling those issues, and to the responsibilities involved. We distinguish:

- a public administration approach in which environmental problems are defined mainly in scientific terms, the government being regarded as the central regulating body, and in which research is chiefly related to the managerial capacities of government. Research is related mainly to the effectiveness of policy (ex post) and the (better) orchestration thereof (ex ante);
- a public management approach in which the environmental problems are defined mainly as complex issues that can only be solved through joint efforts undertaken in inter-organisational relationships. Research here is mainly related to the functioning of policy networks and other consensual management mechanisms and to the circumstances under which they are able to perform even better in terms of both process and product;
- an either sociological or political science approach in which environmental problems are defined mainly as the reason for and the object of societal and political processes of change. Research here is mainly concerned with the social construction of environmental issues, relevant interaction and the question how these processes can be set up as learning processes, i.e. as well-considered and systematic processes of societal change.

As explained, these three approaches were more or less consecutively dominant in the Dutch environmental policy research that evolved since the beginning of the 1980s (see Leroy & Nelissen, 1999, for a more detailed report). The sequence of these approaches can also be described as a gradual widening of the management perspective used: from a strong government-focused view of management, via more inter-organisational management, to more attention being devoted to the societal and political context in which that management must take place. In this sense these views reflect the gradual developments in both the political and the scientific debate on government and governance. The public administration approach dominated the 1980s grosso modo, while the inter-organisational approach has taken the lead since the late 1980s and, as far as we are able to estimate at the moment, has been gradually outshone by the political science approach since the end of the 1990s. This by no means implies that no attention was given to societal change in the 1970s and 1980s, nor that classic evaluative research is no longer being conducted. And yet the shift of predominant approaches is obvious. Moreover, the following will show how much these distinguishable approaches build upon each other's insights and achievements, while simultaneously attempting to correct the alleged shortcomings of the previous period.

11.4 A public administration approach: effectiveness, implementation and orchestration

As already explained, environmental policy originated in a period in which management by a central government was still a plausible prospect. Consequently, the initial environmental policy was very strongly government focused: legal provisions imposed maximum standards, especially regarding the emission of substances which are hazardous to public health. These standards were made operational by virtue of various sector-differentiated permit systems for water, air, waste, soil et cetera. Hence the view that the objectives of environmental policy were mainly scientific issues and that the government should legitimate the results of scientific research in standards. The rest was apparently a question of implementing and enforcing this command and control strategy.

It was this implementation and enforcement that became the central theme of environmental policy research in the early 1980s. Partly because the implementation of various government policies was a popular theme in policy studies (Sabatier, Mazmanian, Wildawsky, Lipsky and others), environmental policy research also focused on this aspect. The key questions were: which of the various policy plans really were being implemented?; to what extent did this damage the effectiveness of the planned policy?; and: which mechanisms could be used to improve the actual implementation of policy? The research related to the implementation of the various laws and permit systems in force at the time focused on environmental problems in general and on water, air and soil pollution in particular. The research in question was inspired by the policy-based and scientifically-relevant question as to the effectiveness of the environmental policy pursued up to then: were the intended objectives regarding water, air and soil quality being met, and if not, why not? The way in which environmental policy was being implemented was believed to play a decisive role in this respect.

One established an implementation deficit in virtually all the research cases involved: policy was changed during its implementation, it was implemented more slowly than planned or incomplete, and enforcement was either defective or inadequate. In some cases the reason for this was to be found in the obscure formulation of the policy plans. More often, though, it was due to the fact that the organisations responsible for the implementation and the target groups of the policy concerned were either unable to accept it unquestioningly, or could not (or would not) comprehend it, or (could not or did not wish to) implement it in conformity with policy. After all, both implementation organisations and target groups had their own concerns and interests, their own motives and possibilities to implement the policy as intended by a central government. The actual implementation could thus differ quite substantially from what had been planned, and consequently could hypothecate the policy's effectiveness.

Studies show that when implementing policy it was particularly the different styles adopted by the different public servants that played a major role. These street-level bureaucrats did not automatically implement policy as it had been planned on the drawing table. After all, the regulations provide these public servants with adequate scope for them to make their own particular interpretation and therefore for policy making too. This scope for policy making is necessary in order to at least make the regulations both feasible for implementation and suitable for specific situations. Nevertheless, it also results in those public servants responsible for the implementation practicing different styles of implementation and enforcement. In a nutshell: they choose a position on a continuum ranging from absolute faithfulness to the formal rules to focusing on the goal and, as a result, choosing for enforcement by way of deterrent or persuasion. Political, organisational and professional factors determine which style of implementation they in fact decide to use. In this respect, research has been conducted into the actual course of the negotiations that took place between civil servants and the representatives of businesses in the permit granting process (cf. Mayntz, 1977). The acceptance of legislation and regulations among civil servants has also been studied as a significant factor of success and failure of policy implementation. If public servants fail to fully accept the policy in question, they will either fail to implement it at all or will not implement it properly. In general, the following (combination of) factors play a major role when implementing environmental policy: the political context, the legal and instrumental pattern, the implementing organisation and the professional attitude of the public servants involved (cf. Wiering, 1999).

The large amount of *ex post* research conducted into implementation, and thus into the effectiveness of policy, almost in itself led among scientists and policy-makers to the question whether that effectiveness could not be boosted *ex ante.* The aspect of orchestration played a particularly crucial role in this respect. The debate that had gradually emerged on management and policy was after all (and not only in environmental policy) focused inter alia on the instruments used by the government (Majone, 1976). In addition to direct regulation – which up to then had been virtually the only management strategy in environmental policy – a call was made for indirect regulation by means of economic and market-conform instruments and for societal regulation by way of communicative instruments. *Ex post* research showed that the different sorts of policy instruments could be either highly effective or less effective, according to the characteristics of the policy problem concerned, the characteristics of the policy field and other policy circumstances. As a consequence, the question was raised whether the effectiveness of instruments could not be estimated *ex ante* when devising policy.

A step-by-step theory of instruments was developed in this context (Klok, 1991). The key question was whether it is possible to forecast which kind of policy instruments is likely to produce the best results, depending on the typical features of the policy field in question (e.g. the number and type of addressees of policy, the implementing organisation, etc.) and on the intended processes of behavioural change. The instruments theory is thus perfectly in line with the policy debate conducted at that time on the diversification of policy instruments and the introduction of different sets of instruments for different constituent fields of environmental policy. The instruments' theory and its application in various fields of environmental policy came up with some interesting insights into the constituent processes of policy implementation, both in terms of the governmental organisations involved and the groups for whom the policy was intended. The actual working of a number of agreements, the working of several levies and subsidies, and the scope and effectiveness of information, were all studied (Vermeulen, 1992). Worthy of note in this respect is that while calls were and are made both on national and international level for the implementation of economic instruments, so far little empirical research has been conducted into the actual effects of those levies and subsidies and the associated instruments. Hitherto, therefore, the argument is not backed up by empirical evidence.

It became quite clear from the many research projects that made use of the instruments theory – sometimes explicit and sometimes implicit – that the main aspect in orchestrating policy and influencing the target groups was a question of highly complex processes that cannot be depicted in mechanistic diagrams of 'stimulus response' or 'a change in policy instruments and behavioural change'. Research into the instruments tended to make the rationalistic points of departure problematic, and therefore brought them into perspective. Studies were carried out into the effectiveness of subsidies in several fields of environmental policy (including heavy metals, PCBs, 'low noise' lorries, and the storage of manure). A significant conclusion drawn from this research was that subsidies have little ability to penetrate the internal decision-making processes of businesses and organisations, and that, in most cases, they give some support to existing plans for more environmentally friendly investments.

11.5 Governance in context: policy networks as an organisation strategy

As already set out in the foregoing, the instruments theory itself gradually made its own points of departure more problematic. This also applied with regard to the initially strong focus on government instruments. Both policy research insights and considerations from policy practice led to a higher level of interest among the non-governmental actors involved in the policy. In the sense of policy studies, the more the instruments theory evolved, the clearer it became that 'reliable orchestration', especially of the field variables, i.e. circumstances, organisations and power relations, was dependent on elements outside the field of government. The relative autonomous actions of target groups were more determinative for the outcomes of policy than the public administrations' rationality of government intervention.

Running parallel with this growing insight, the concept of 'internalisation' appeared in the formulation and practice of environmental policy in the mid-1980s. This meant that government should not force businesses and citizens to behave in a more environmentally responsible fashion, but that the environment ought to be an internalised cultural element, inducing that environmentally friendly behaviour more or less automatically. In terms of politics this brought up the argument to reorganise the division of tasks among state, market and society. The concept of 'internalisation' soon became very popular, simply because it was in line with the pleas for the reallocation and reorganisation of responsibilities among the state, the market and civil society argued in the wider sense.

The programmatic concept of 'internalisation' went much further than simply a shift in type of instrument, in this case from direct to indirect and society regulating. The term also concerned the division of tasks and the organisation of environmental policy in particular. Substance and form was given to this concept in two separate fields in particular: in the target groups policy and region-oriented policy. Both of these policy strategies have repeatedly – often 'by request' – been the subject of policy research. The results of that research have helped to support its further development.

The core of the target groups policy, which apart from the Netherlands has since been given shape and form elsewhere in Europe and at EU level itself, is to increase the own responsibility of market parties in particular. The idea is that the societal, mainly economic activities that cause a high level of environmental load (agriculture, industry and a few smaller groups in industry, traffic etc.), contribute towards environmental policy in consultation with the government, and are even held co-responsible (see also Chapter 4). The target group can see to the practical implementation of objectives agreed with government on the reduction of emissions, the use of energy etc., via the relevant sector organisations. This implies less government involvement and more autonomy for trade and industry; the government consequently has less responsibility for implementation and enforcement. The target groups policy has meanwhile led to the institutionalisation of regular consultation between government and the market parties, and to the realisation of a series of covenants or similar agreements. In most of the economic sectors involved this has also led to the setting up of sector associated organisations for environmental expertise and advice. While this does not lead to the actual replacement of direct regulation via legislation and regulations, it does lead to it being supplemented with self-regulation in the sector concerned. A variety of standards and systems for in-house environmental protection have been developed, both internationally and elsewhere in Europe in the wake of this shift in responsibilities. Implementation, certification and accreditation have in part been contracted out to specialised private companies.

Consequently a shift in tasks and responsibilities has come about, particularly from government agencies to market parties in the target groups policy. Representatives of civil society hardly play a role of any significance in this respect. It would even seem – if only because of the various bilateral agreements, which are still only confirmed in administrative procedures and governmental decrees – that there is less rather than more 'third party' influence on the decision-making process.

Region-specific environmental policy is another innovative strategy of environmental policy that dates from the late 1980s (see also Chapter 6). This is typical of the new role relations between government, market and society. It was shaped predominantly in projects that were intended for the purpose of converting a joint formulation of the problems into a broadly based, area-specific policy. This was accomplished on the basis of a largely organised consultation with the representatives of a variety of local and regional groups and organisations (in the agricultural sector, housing, industry, tourism, nature conservation). The government therefore does not impose its generic policy but rather negotiates on the question how these policies can be adapted to the specific regional situation with local authorities, stakeholder interest groups and citizens. The objective of these negotiation is (a) to coordinate and integrate policy efforts on a number of policy fields (e.g. agriculture, environment, spatial planning, traffic, tourism etc.) on a regional basis, and (b) to create sufficient support and legitimacy among those concerned to ensure that the policy also stands in the implementation stage. Region-oriented policy thus attempts to provide a combined answer to the relevant compartmentalisation of policy fields and the lack of implementation of too generic and consequently locally rejected policy. It also attempts to benefit from the experiences and insights regarding the implementation and orchestration of policy. This form of policy application is referred to in international literature in the field of policy studies as inter-organisational policy making, policy mediation, negotiated decision making, cooperative policy making and inter-organisational implementation (Glasbergen, 1995 and 1998).

Research conducted in policy studies into region-oriented policy devotes particular attention to the creation and the effect of networks of teams of organisations or 'policy networks'. Initially, the emphasis was on the necessity and the actual functioning of policy networks in relation to these projects. The necessity of policy networks was argued mainly on the basis of government's limited managerial capacities and the reciprocal dependency of those actors who were able to solve a part of the problem. This pulling together of the actors involved in policy networks would therefore seem to be inevitable for the purpose of generating the managerial ability needed to tackle the problem. In addition to the subject of analysis, the activation and building up of networks then also becomes a management strategy.

Studies conducted into the region-oriented approach showed that this strategy usually succeeded in getting the actors involved in a policy process to join forces and pull together, but that this by no means guaranteed the quality and effectiveness of the actual policy content (De Jong, 1999). A smoothly running process is no guarantee for a good policy product. Others conducted their studies into region-oriented policy less on the basis of a process-based, public administration context and devoted more attention to the political context. They typified region-oriented policy as a part of a more encompassing devolution, albeit it in different guises, of environmental policy throughout Europe. Partly as a result of the fragmented way in which environmental policy was institutionalised in the 1970s and 1980s its coordination with other fields of policy and its adaptation to various regions were problematic. By, for instance, attempting to reconcile nature and the agriculture sector, housing and recreation and other sectoral interests on the basis of projects, and by involving the various stakeholders in the preparatory work, the decision-making process and the implementation, it is quite possible that area-oriented policy can offer a remedy. At the same time, however, this approach runs the risk that various political debates that cannot be resolved at the central level are simply pushed onto the regional level. The mediation techniques used to tackle the environmental problems in the vicinity of numerous airports, from London Heathrow to Vienna and from Amsterdam Schiphol to Zürich, show both the benefits and the risks of this approach.

Nevertheless, both the target groups policy and region-oriented policy illustrate the shift in dominant approaches and strategies in environmental policy. This policy, in comparison with that pursued in the 1970s and 1980s, is dominated far less by government, and is increasingly being realised through the complex efforts of many actors with a variety of interests. New forms of management emerge in this context. Some researchers lay the emphasis on analysing the circumstances under which these new forms of management can arise, and on the resulting consequences for the relationships of power among the actors involved and the actual content of policy. Others investigate how this joint effort by representatives of government, market and civil society can be put to use to achieve a higher quality of both policy process and policy content. Instead of a debate on the 'most effective instruments', from the second half of the 1990s the debate turned mainly onto 'good teamwork', or better: onto adequate processes and methods for the activation, the organisation and the structuring of co-operation between state actors, market agencies and civil society representatives. These attempts have to be interpreted within the context of the changing interrelations and the shared responsibilities between state, the market and civil society.

11.6 The political construction of discourses and the political context: policy learning?

A major difference between the two approaches discussed above is their amount of faith in the government's managerial capacities and the extent to which they assume that, to effectively solve the problem, it is essential to secure the involvement and efforts of several, if not all parties involved. One important consistent element is that both approaches place strong emphasis on the strategic, instrumental and organisational aspects of policy, and attach relatively less importance to the actual policy content. In both cases, the concern is not so much to investigate the policy objectives, but to find out the strategically best way to achieve them.

The latter is, however, only partly correct: the network approach has helped to show how important convergent problem definitions are as a precondition for the parties concerned to enter into discussion and to formulate common strategies for solving them. Yet the most important incentive for devoting attention to the contextual, discursive aspects of activities or actual policy content still comes from the social constructivist perspective that is (again) in fashion in the social sciences. It is important for policy studies in particular that more attention is devoted to the processes of naming and framing policy problems, to the importance of policy concepts as the instigators of and lubricant for policy processes, in short, to the role of policy discourses.

These processes of naming and framing are by no means non-committal: giving societal and political meaning to the problems results in institutionalised patterns of behaviour, in organisations and rules, in structures. This also applies to environmental policy. Environmental problems and environmental policy problems are analysed in a socio-constructivist approach mainly as socially defined problems. Such problem definitions give rise to the formation of coalitions and divides among the actors involved (cf. Hannigan, 1995). Moreover, the problem definitions and associated coalitions are apparently reasonably stable and relatively inert (Sabatier & Jenkins-Smith, 1993). This contributes towards a blockage in policy or a lack of dynamism, obviously partly caused by the conscious and subconscious strategic behaviour of those concerned. On the other hand certain organisational traditions, for instance a certain division of tasks between governmental organisations, lead to the continuation of certain, sometimes no longer adequate problem definitions and strategies to solve them. Discourses are therefore selectively associated with specific organisational contexts; conversely, certain organisational contexts lead to certain dominant discourses (Eberg, 1997).

International comparative environmental policy research has recently produced some interesting findings from this perspective. It is the connection between the 'hard' institutional differences and the 'soft' problem definitions that helps to explain why environmental policy has developed so differently in the different countries, and is still organised differently today. An international comparative study clearly demonstrates how acidification is framed and named in a quite different way in Great Britain and in the Netherlands; this resulted in a completely different development of policy (Boehmer-Christiansen & Skea, 1991). These dominant policy theories and policy trends are obviously built on previous perceptions of the problem and its institutionalised expression in a variety of structures, strategies and instruments. For instance, a comparison was made between the Dutch and the Bavarian waste disposal policy (Eberg, 1997). This study used the concepts formulated by Sabatier and gave attention not only to the dominant problem definitions and policy concepts, but also to the stability and dynamism of the coalitions based thereupon. These aspects explained the likelihood of opportunities for 'policy learning', in other words, for well thought out policy innovations. This reasoning and these findings led to subsequent questions concerning transnational and international environmental policy, and to questions about the different ways in which, for instance, various nation-states implement European directives (Haverland, 1998; Arts & Van der Zouwen, 1999).

Recent studies have also shown quite clearly that this approach is also of interest for the analysis of 'domestic' environmental policy. Verbeeten (1999) analysed decades of public and political debate on the Wadden Sea and showed how difficult it is to reconcile the diverse perceptions of the problems in that field. She also showed how difficult it is to use the different perceptions as a basis for more or less identical views to reach a solution. In such circumstances the 'learning' element is complicated. Pestman (2001) points out the strength of the actor coalition that has supported the slogan 'Nederland distributieland' *(The Netherlands as the Gateway to Europe)* since the mid-1980s. This large and particularly powerful coalition was reasonably unanimous in its support of the expansion of Schiphol Airport, of the construction of the second 'Maasvlakte' (an industrial estate near the port of Rotterdam), the 'Betuweroute' (a new railway line connecting the port of Rotterdam with Germany), the TGV and other plans. Local protests, and even the objections put forward by what on paper was a heavyweight coalition of provincial and municipal authorities, environmentalist groups and scientists, were still unable to change that unanimity.

11.7 A retrospective of environmental policy research

As already pointed out, policy studies are a young and not precisely defined collection of disciplines with many different lines of approach. The status attached to policy studies differs greatly from one country to the next, and this certainly applies with regard to the amount of attention they devote to the environment. In turn, environmental policy is a dynamic and heterogeneous field of policy. This combination can lead to no other conclusion than that environmental policy studies research is highly diverse, both in terms of the theoretical approaches used and the empirical fields described. It is therefore difficult to make an assessment of this diversity. Nevertheless, the overview does lead to a few conclusions that at least apply to the Netherlands.

Firstly, it has been made quite clear that with its successive dominant approaches and research questions, environmental policy research is to a large extent influenced by the shifting needs for knowledge regarding the evolving environmental policy. Conversely, it becomes perceptible that the research itself is influencing certain aspects of policy. Throughout the past two decades there has been a strong link between the development of environmental policy and the development of environmental policy research. That they have an influence on one another is in itself good: policymakers are able to question scientists about their needs for knowledge and, in turn, the insights of policy researchers apparently help to further policy making. However, there is also a risk attached to this tight-knit interaction: if questions put to policy studies are based too heavily on what today is the dominant subject of policy (inspired by the politically strongest actor's need for knowledge), then other scientific issues threaten to become relatively neglected. This applies all the more so if the researchers are more or less financially dependent on contract research. This is the case with regard to the environmental policy research in the Netherlands, as it is funded for about 60% by contract research, a figure that is way above the average in other disciplines (Leroy, 1995).

Certain elements imply the risk that two other sorts of demands for research become relatively neglected: (a) the demand for more fundamental research, and (b) the demands for research from the financially or otherwise weaker groups in society. The former includes, for instance, social science research conducted into the relationship between structural societal developments, such as individualisation and globalisation, and the associated relatively marginal steering role of environmental policy, or research into the institutional changes required for sustainable development. The latter, for example, covers requests for research made by citizens and local groups confronted with plans that threaten the environment (both in the Netherlands and elsewhere in the world). The weak position, also with regard to scientific support for inhabitants and environmentalist groups, stands out sharply against the overwhelming amount of policy-supporting research which their opponents have in these matters at their disposal.

Secondly, the foregoing clearly shows that environmental policy research, such as that conducted in policy studies in general, also has a twofold objective: to conduct research both *into* and *for* policy. And yet one still has the impression that, partly because of the close cooperation that exists between policymakers, the managerial objectives generally win from the analytical ambitions. In other words: environmental policy research has a high 'public administration content' and is inspired far less by sociological and political science motives. We should precise this latter conclusion: for the sake of brevity the above overview has made no mention of, for instance, political science research into the role and influence of the environmentalist movement in both the national and international dimension, or of sociological research into the changes in consumer patterns and the lifestyle of citizens. Nevertheless, even this qualification leaves us with the conclusion that environmental policy research has a high public administration content. It involves investigations into processes (especially implementation and orchestration), organisation and effects (for the latter especially: effectiveness; much less: side effects), and much less attention is focused on the content and context of policy. In this sense, the recently renewed attention focused on policy content and policy context is a welcome rectification. However, both the contribution and the added value it provides, are still difficult to estimate precisely.

References

Arts, B. & M. van der Zouwen (1999) Policy Arrangements in Nature Conservation: The Case of the Netherlands in the European Context, paper presented at the ERP conference *Environmental Policy in Europe: Visions for the New Millennium,* September 13-14, London.

Boehmer-Christiansen, S. & J. Skea (1991) *Acid Politics. Environmental and Energy Policies in Britain and Germany,* Belhaven Press, London.

De Jong, D. (1999) *Tussen natuurontwikkeling en Landschaftsschutz. Sociaal-cognitieve configuraties in het grensoverschrijdende natuurbeleid,* Eburon Publishers, Nijmegen/Delft.

Dinkelman, G.H. (1995) *Verzuring en broeikaseffect. De wisselwerking tussen problemen en oplossingen in het Nederlandse luchtverontreinigingsbeleid 1970-1994,* Van Arkel, Utrecht.

Eberg, J. (1997) *Waste Policy and Learning. Policy Dynamics of Waste Management and Waste Incineration in the Netherlands and Bavaria,* International Books, Utrecht.

Glasbergen, P. (ed.) (1995) *Managing Environmental Disputes. Network Management as an Alternative,* Kluwer Academic Publishers, Dordrecht.

Glasbergen, P. (ed.) (1998) *Co-operative Environmental Governance. Public-Private Agreements as Policy Strategy,* Kluwer Academic Publishers, Dordrecht.

Hannigan, J. (1995) *Environmental Sociology,* Routledge, London/New York.

Haverland, M. (1998) *National Autonomy, European Integration and the Politics of Packaging Waste,* Thela Thesis Publishers, Utrecht.

Klok, P.J. (1991) *Een instrumententheorie voor milieubeleid,* Twente University Press, Enschede.

Leroy, P. (1995) *Environmental Science as a Vocation,* Department of Environmental Policy Sciences, Nijmegen.

Leroy P. & N. Nelissen (1999) *Social and Political Sciences of the Environment. Three Decades of Research in the Netherlands,* International Books, Utrecht.

Liefferink, D. (1996) *Environment and the Nation State. The Netherlands, the European Union and Acid Rain,* Manchester University Press, Manchester.

Majone, G. (1976) 'Choice Among Policy Instruments for Pollution Control', *Policy Analysis,* pp. 589-613.

Mayntz, R. (1977) *Vollzugsprobleme der Umweltpolitik,* Kohlhammer, Stuttgart/Wiesbaden.

Pestman, P. (2001) *In het spoor van de Betuweroute: mobilisatie, besluitvorming en institutionalisering rond een groot infrastructureel project,* Rozenberg Publishers, Nijmegen/Amsterdam.

Sabatier, P.A. & J.C. Jenkins-Smith (1993) *Policy Change and Learning. An Advocacy Coalition Approach,* Westview Press, Boulder/San Francisco.

Smits J. & A. Ringeling (1997) 'Verweven wetenschap. Onderzoek van en voor milieubeleid', in: P. Leroy (red.) *Verbondenheid en samenhang. 10 essays bij het afscheid van Klaas Bouwer,* Department of Environmental Policy Sciences, Nijmegen, pp. 195-217.

Van Tatenhove, J., B. Arts & P. Leroy (eds.) (2000) *Political Modernisation and the Environment,* Kluwer Academic Publishers, Dordrecht.

Vermeulen, W. (1992) *De vervuiler betaald? Onderzoek naar de werking van subsidies op vier deelterreinen van het milieubeleid,* Van Arkel, Utrecht.

Verbeeten, T.C.M. (1999) *Wijs met de Waddenzee? Een onderzoek naar leerprocessen,* Thela Thesis, Utrecht/Amsterdam.

Weale, A. et al. (2000) *Environmental Governance in Europe,* Oxford University Press, Oxford.

Wiering, M. (1999) *Controleurs in context.Handhaving van mestwetgeving in Nederland en Vlaanderen,* Koninklijke Vermande, Lelystad.

Chapter 12

NEW DIRECTIONS IN ENVIRONMENTAL POLITICS

Concluding remarks

Peter P.J. Driessen and Pieter Glasbergen

12.1 Introduction

In the previous chapters we have take a look at the paradigm shift that has taken place in Dutch environmental policy over the past few years from a variety of angles. The changes observed relate to strategic thinking in terms of the actual content of environmental policy and the procedures that were adopted. As has been argued by many of the authors in this book, the changes concerned must be seen within the shifting relations between state, market and civil society.

The term 'paradigm shift' was originated by Kuhn, a historian of science. In order for us to really plumb the depths of this expression it is appropriate that we first take a look at how he used the concept. In his now well-known book 'The Structure of Scientific Revolutions' (1962) Kuhn presents a theory of the scientific progress within which paradigms play an important role. The term 'paradigm' is indicative of one of the main streams of scientific thinking. A paradigm shift therefore relates to a radical change in this mainstream of science. According to Kuhn, the traditional idea of steady growth in our scientific knowledge by virtue of the accumulation of knowledge is misleading (Kuhn, 1962, p. 2). On the other hand, there is talk of progress by virtue of revolutionary breaks.

For according to Kuhn, if we examine any branch of science we find that there are certain fundamental theories and conceptions which hold steady for very long periods. These settled views of the world become so fundamental to people's whole conceptual scheme of reality that it is

P.P.J. Driessen and P. Glasbergen (eds.), Greening Society, 245–262.

extremely difficult for most people to think of reality as being different. It concerns, according to Kuhn, a period of 'normal science', in which research in that field is performed completely within the framework of a paradigm. That period is characterised by intense stability and concurrence. Normal science is seen by Kuhn as research "that is firmly based upon one or more past scientific achievements, achievements that some particular scientific community acknowledges for a time as supplying the foundation for its further practice. Today, such achievements are recounted by science textbooks, elementary and advanced. The achievements are sufficiently unprecedented to attract an enduring group of adherents away from competing modes of scientific activity. Simultaneously, it is sufficiently open-ended to leave all sorts of problems for the redefined group of practitioners to resolve" (Kuhn, 1962, p. 10).

However, a period of normal science can also turn into a period of crisis and lead to a revolution in which the paradigm itself, or parts thereof, are at stake. During the period of normal science, in which a given paradigm prevails, scientists are often aware of evidence which the prevailing paradigm is apparently unable to explain satisfactorily. Most scientists do not allow this counter evidence to unseat their adherence to the paradigm. Conversely, the main explorers in the history of science have shown an enormous interest in these irregularities and used them as the basis for new theories. These led to crises in science and to furthering the realisation of alternative paradigms. In Kuhn's words: a paradigm shift, a revolutionary break in the way of thinking, because the way in which certain phenomena are observed and understood are upturned by a new theoretical perspective. In the revolutionary period we see the defendants of the old paradigm opposing the supporters of the new paradigm. One well-known example of this was when Copernicus overturned the ancient scientific belief that the world was flat and that it was the centre of the universe. The triumph of a new paradigm is again followed by a (generally long) period of normal science.

It is probably taking things too far to use the concept of the paradigm shift – which was originally introduced to indicate major changes in the philosophy of science – to analyse major changes in the way in which we deal with environmental issues. Developments in science as analysed by Kuhn, often relating to very lengthy periods (sometimes centuries), can hardly be compared with developments in environmental politics that relate to only a few decades. Moreover, Kuhn's view concerns the development, the establishment and the overturn of *scientific theories*. Environmental politics is usually based on scientific theories to only a very limited extent and mainly on *suppositions* as to the effectiveness of societal intervention strategies. And yet we have still chosen to use the term 'paradigm shift' in this book; not so much in the literal sense – as used by Kuhn – but more in the symbolic sense: as an indication of the fundamental, almost revolutionary changes in the way we think about how radical societal changes can be brought about which lead towards a more sustainable society. These changes emerge from an entirely different way of looking at social reality, i.e. at societal processes and how they can be influenced.

In this chapter we draw up the balance of the paradigm shift that has taken place in Dutch environmental policy. In the following we shall discuss the most important elements of the paradigm shift, partly on the basis of that which has already been argued in the previous chapters. Subsequently, we take a look at the driving forces behind those shifts. In other words: what factors played a role in initiating advances in our views on the actual content of environmental policy and in the policy strategies to be used? One theme which certainly may not be overlooked here is the effect of the paradigm shift. This book has looked at the environmental issues from different angles: from the perspective of production, consumption, rural and urban areas. What are the opportunities and limitations that present themselves in the new approach to these issues? We close this chapter by taking a look at the value of the new paradigm.

12.2 The characteristics of the 'old' paradigm

The genesis of environmental policy in the Netherlands and in other western nations can be traced back to about 1970. Even before then had initiatives been taken in the direction of environmental policy (for instance in the form of noise prevention) yet they did not have the all-embracing and radical characteristics we see in today's policies. One major impulse for the birth of contemporary environmental policy were the crises that occurred: early in the seventies significant cases of soil pollution, water pollution and air pollution were seen in the Netherlands and elsewhere in Europe. All of which were associated with risks for public health. Additionally, waste problems and 'nuisance' problems (noise and stench) made significant contributions towards the increasing amount of attention given to the quality of the environment. The government reacted by systematically developing and implementing a variety of measures. This was analysed by Van Tatenhove & Goverde elsewhere in this book (see Chapter 3). Originally, the status of environmental policy within the totality of government policy was indistinct. Its position had to be won and perpetuated. Simultaneously, expectations were high as to what could be achieved with environmental policy in the short term. This was also in line with the political and ideological culture of the day. In the seventies – the heyday of the welfare state – there was enormous faith in government and in the makeability of society. Heavy government intervention to improve the quality of the physical environment fitted in well with this view. The first sizeable policy document on environmental policy – the Urgency Memorandum (1972) – exuded an atmosphere of both serious crisis and faith in the solving thereof within the foreseeable future. Also the institutionalisation of environmental policy in the form of strategies, organisations, the distribution of responsibilities, instruments, etc., was in line with the views of the welfare state: the government took the leading role in environmental policy and only saw itself obliged to inform the market, non-governmental organisation and citizens of its policy plans.

This view on environmental policy can be seen as a 'paradigm'. As has already been argued in the foregoing, this is not a paradigm in the scientific sense (i.e. a scientific theory), but in fact the totality of assumptions on which the policy is based. It is therefore more of a pre-scientific theory (see Leroy, 1994, p. 36). The assumptions relate to cause-effect relationships, goal-means relationships (impact models) and current values and standards. They are assumptions that play a role among public servants and administrators when moulding and implementing environmental policy. In other words, hidden behind this paradigm is a view as to the environmental issues and how they should be resolved.

The main paradigm in the Dutch environmental policy of the nineteen seventies was based on the following assumptions (see also Glasbergen, 1994, pp. 28-29; Snellen, 1987):

- Environmental issues are mainly local and regional problems which are closely linked with the issue of public health.

- It is the government that should have the leadership in the process of societal change. It is the government that sets out environmental policy and should be responsible for its implementation. In other words: the government acts as the subject as opposed to society as the object.

- The input of scientific knowledge is required in environmental processes in order to improve rationality and therefore the quality of policy. Scientific research thus contributes towards knowledge about environmental issues and their definition, and towards defining and operationalising environmental goals. Natural scientific research in particular should play a major role in the actual formulation of environmental goals.

- Environmental problems can be solved. The policy is formulated in such a way that the environmental goals can be realised within a period of approximately ten years.

- The environmental goals are raised above other social objectives and are seen as standards that must be met by the market, NGOs and citizens.

- The environmental goals are used as the basis for searching for instruments that bring us closer to realising those goals. Technical solutions are essential for each sector (water, soil, air, waste and nuisance problems). Technological scientific research is thus stimulated. Additionally, legal measures in the form of orders and prohibitions also help to achieve the goals.

- If the right instruments are chosen, businesses, non-governmental organisations and citizens will respond in the way government intends them to respond, and the environmental goals will be achieved.

- Communication between government and the societal actors is particularly in the form of knowledge and information transfer and is intended to convince others that the best solution has been chosen.

This was the paradigm that determined Dutch environmental policy in the nineteen seventies and early eighties. A similar paradigm was also to be seen at that time in other liberal democratic countries. These assumptions must be seen with subtle distinctions. In practice, the paradigm existed in several variations. Yet the main theme was always the government's hierarchical position, the dominant role played by scientific and technological knowledge, plus the rigid goal-means philosophy. It was also typical – in line with Kuhn's thesis – that it was always possible to solve the problems in governance within the existing paradigm. When it became apparent that environmental policy was unable to tackle the problems successfully, the steadfast reaction was: formulate the environmental goals more critically, underpin the goals even better with scientific research, increase the policy efforts by a better and more intensive use of policy instruments, and finally, impose more strict and intensive control on the actual implementation of the policy.

In this period, the development of environmental policy and its implementation was frequently the subject of scientific research. This, as is apparent from Leroy & Nelissen's contribution to this book (see Chapter 11), led to studies into the practice of policy implementation and policy enforcement, into the actual effects and the effectiveness of environmental policy, and into the choice and use of policy instruments (in this particular case: legal and economic instruments and instruments of communication). In hindsight we can state that this scientific research also made a substantial contribution towards supporting the existing paradigm.

12.3 The development of a new paradigm

The paradigm described in the foregoing, on which environmental policy had been based for many years, has without doubt been a success. In many countries, especially in the Netherlands, a reduction in emissions has been achieved which – to all intents and purposes – has reversed the immediate health-threatening areas of pollution. Nevertheless, the paradigm has shown wear and tear along the road. Scientists and political figures alike became aware that some of the essential components of the paradigm were no longer viable. In this connection there were three developments of particular significance.

First and foremost, *major shortcomings* were observed *in the implementation of environmental policy.* While various elements of the policy pursued were apparently successful, several important components thereof were failing in the operational sense. Environmental policy was fragmented in terms of content, organisation and procedures, and thus some of the problems were not so much solved but rather passed on from the national government to regional and local authorities and from the one societal sector to the other. Moreover, many environmental issues were apparently more difficult and slower to solve than had been expected. The classic reaction to the failing policy – the more specific formulation of goals, the better use of instruments and more strict control on implementation – soon appeared to bear insufficient fruit. New problems of a very complex nature were gradually added, the pollution of channel beds for instance, global warming and the depletion of the ozone layer. Whereas in the nineteen seventies and early eighties it was often a case of local, easily recognisable problems of pollution, more diffuse problems were gradually observed, problems that were linked to complex production and consumption chains. To solve these problems, specific target groups needed to be addressed and convinced of the necessity of far-reaching changes in production and consumer patterns. The traditional way of taking action soon appeared unsuitable in this respect since they offered little basis for the policy to be pursued.

Secondly, changes took place *in the strategic points of departure of environmental policy,* both in the national and international dimension. Policy became more focused on the aim to achieve a sustainable society, particularly under the influence of the Brundtland Commission's views ('Our Common Future'; OECD, 1987). The introduction of the term 'sustainability' shed a different light on environmental issues. A balance would have to be found between environmental qualities on the one hand, and societal and economic developments on the other; a balance acceptable both in the short and the long term. In other words the problems had been widened: sustainability was now seen as a development issue concerned with the future, an issue for which broadly-based public considerations were essential. However, despite a growing awareness of the environment among the citizens, the government's good intentions, and the systematic structure of environmental policy, the general opinion was that far-reaching transitions in the direction of a sustainable society with the help of the old paradigm would probably be impossible to achieve. After all, the old paradigm was not based on taking all considerations into account (i.e. searching for a balance), but on the unilateral setting of demands (in the form of environmental standards) on the different sectors of society. The aim towards sustainability would therefore need to be linked up with another strategy: the involvement of all parties that might possibly be able to influence that sustainability. In other words: a different method of working, focused on collaborations between central government, local authorities, the market, NGOs and citizens.

Thirdly, *the developments in government policy in the general sense* gave impetus to the paradigm shift in environmental policy. The welfare state was apparently in need of radical reform. The government was unable to sufficiently fulfil its role as the central governing organisation in practice. Society was less makeable than the government had anticipated. The welfare state with its extensive tasks and responsibilities for the government was neither non-implementable nor affordable (Godfroij & Nelissen, 1993). Conversely, the government would have to take a step down, needing to take upon itself a more encouraging role in civil society and would need to give that society more opportunity to take initiatives. Over the course of the nineteen eighties this led to the hiving off of government tasks, decentralisation and deregulation, and also to a reduction in public expenditure. Simultaneously, it also implied an ever-increasing interweave between government, the market and non-governmental organisations. The resulting complex relationship patterns and the distribution of power and influence gave stimulus to the views on network management as has been seen in various sectors of policy since about 1990.

In the mid-eighties environmental policy was strategically revitalised; this was explained by Van Tatenhove & Goverde in Chapter 3. This renewal related to both the content (resulting among other things in internalisation, target group policy and area-oriented policies) and the procedures adopted (forms of network management; see Chapter 2). A new paradigm emerged 'en route' based on the following assumptions:

- Environmental issues are seen as complex problems that occur on all the different spatial levels (local, regional, national, fluvial, continental and global). Moreover, they are not only linked to the issue of public health but also to issues concerned with the preservation of natural resources and biodiversity.

- The approach taken to deal with environmental issues is the responsibility of government *and* society. Central government, local authorities, the market and non-governmental organisations are related horizontally and all are in possession of instruments to realise the environmental goals.

- Scientific knowledge is an important factor in environmental processes. At the same time, in addition to scientific insight, attention must also be given to political views, empirical knowledge and the experiences of non-experts. Input from the scientific world must not remain limited to scientific knowledge, but to knowledge regarding environmental issues available in fields of social science, and potential solutions to these issues must also be brought into the policy-making processes.

- A successful approach to environmental issues calls for patience; solutions to complex issues are only expected in the longer term (above 25 years).

- Environmental goals are placed in a wider development perspective in which the social and economic goals are given equal importance. Moreover, in the development perspective there is the aim to create new opportunities for those societal interests that must be subjected to a process of re-orientation because of the environmental goals.

- The approach taken towards environmental issues must go hand in hand with radical 'transitions'. Not only technological innovations are necessary to stimulate a more sustainable society but also societal changes, i.e. changes in how individuals and organisations behave, institutional changes, socio-economic changes and changes in the political culture.

- In order to solve the problems, structured processes of interaction with, and communication among the most involved actors must be started up. These are actors associated with the government, the market and/or non-governmental organisations.

- A successful approach towards environmental issues can only be realised if consensus has been achieved between the parties most involved in the issue.

- Communication between government, the market and NGOs is bilateral by nature. Information, views, interests and perspectives are exchanged in order to achieve agreement on the problem and the most desirable and feasible solution.

The new paradigm developed gradually in the Netherlands. After several initiatives had been taken towards a rejuvenation of environmental policy in the mid-eighties (see Chapter 3 and the societalisation and regionalisation strategies discussed therein) network management started to take shape in the nineteen nineties in a more well-considered and structured manner. Today, this form of governance is applied in a variety of gradations. The different variations relate mainly to the parties involved, the extent to which other perceptions of the problem and the options to solve it are allowed, and the methods used to stimulate consensus among the parties involved.

One of the main features of the new paradigm is the so-called 'pluricentric perspective' (Teisman, 1992). Environmental policy is developed and implemented within networks, as has already been stated by Goverde & Nelissen in Chapter 2. If it is the intention to achieve something, then it is essential to form a network. The parties that make up that network are dependent on one another. Their interests and needs cannot be achieved without one another's help. According to the new paradigm, policy can only be reached and implemented by bringing together these parties with their different interests, goals and means. However, these parties will not wish to communicate with one another if the problems and solutions are looked at from only one point of view. In other words, there must be room for different views and perspectives in policy-making processes. In practice this generally means 'a widening of the problem': environmental issues are linked then to other societal issues. This implies that the solution will frequently have a wider, more integral character. It is also striking to note that uncertainty as to the outcomes of the process are accepted in the new paradigm. Within certain boundaries, especially with regard to the time frame, environmental goals are negotiable. Solutions are not established in advance but developed in a process of negotiation. The parties involved will need to come together and learn how to realise outcomes that both bring the environmental goals closer and do justice to other societal interests. Finally, it is not a question of a single government taking a decision, but decision-making in networks is characterised by a series of decisions taken by what can sometimes be a changing number of actors.

Here too are the problems connected with the implementation of environmental policy also solved within the dominant paradigm. Impasses in the preparation and implementation of policy are broken through by mobilising or involving other actors in the process, by widening the problem, improving the communication processes (in order to enable the parties to learn more from one another) and by supervising the negotiations better between the parties concerned.

The development of this paradigm was also 'guided' by scientific research. This mainly concerned research into policy networks (applied on various forms of interactive governance, such as the target group policy and regional-oriented policy). Research was also conducted into the discursive nature of policy processes and the naming and framing of policy problems and the associated solutions (see Chapter 11).

Policy development in the international arena also helped to develop and establish the new paradigm. International environmental policy became more and more important during the course of the nineteen nineties. Even more so than at the national level, hierarchical governance is virtually incapable of achieving the intended environmental effects at international level, particularly in the longer term. Arts, Dieperink & Liefferink state in Chapter 10 that international environmental policy is a process that is seen increasingly among the numerous actors on the different geographical and administrative levels. Policy is realised through complex teamwork processes between international institutions, NGOs, national government, regional and local authorities and private actors. It has become a question of multi-actor governance and multi-level governance. This approach often has consequences for the choice of instruments. For instance, innovative mechanisms were developed to achieve a reduction in greenhouse gases within the framework of the climate change policy. Agreements were reached in the Kyoto Protocol on Joint Implementation (JI) and the Clean Development Mechanism (CDM); instruments based specifically on cooperative management. Also the 'sustainable development treaties' entered into between the Netherlands and countries such as Costa Rica, Benin and Bhutan are examples of cooperative management. Success in the international arena with the new paradigm can imply a further stimulus for the acceptance of new methods of governance at the national level.

12.4 The new paradigm and a different view on democracy

The new paradigm was regarded for many years in the instrumental sense only. It was a new way to realise societal goals. Network management was essential for its effectiveness. The goal of the teamwork between public and private actors is to realise policy either better or more quickly. Other parties often have instruments that can promote successful implementation. By pulling together, policy support is created among all parties concerned and thus preventing the risk of legal resistance in the implementation stage and, in turn, delays.

The new governance paradigm was also motivated and stimulated in the course of the nineteen nineties on the basis of a new perspective on democracy. In the former paradigm it was the government that made decisions conform the rules of representative democracy and the citizens and private organisations that influenced those decisions by way of formal rules on participation. The assumption was that the democratic content of the decision-making process would be increased in the new governance paradigm. 'Openness' played a significant role in this sense. A decision-making process is open if all parties with a stake in an environmental issue are involved when that issue is being dealt with, and that that the process itself is open for inspection by both those concerned and outsiders. In other words, more parties are able to participate directly in the policy-making process and subsequently in the actual decision-making and the ultimate implementation. Moreover, in this interactive policy-making process there must be a certain equality between all the parties involved. Differences in power and the availability of resources in the form of knowledge, means and expertise must be suppressed. Central in the democratic perspective on the new governance paradigm is therefore the higher level of influence businesses, non-governmental organisations and citizens have on the decision-making process.

This new democratic perspective has been dealt with twice in this book. Leroy & Van Tatenhove set out in Chapter 8 that since the nineteen seventies environmentalist organisations have protested against the adverse environmental impact of numerous government decisions, especially against the way in which those decisions were taken, i.e. the lack of opportunities to participate in the decision-making process. They argue for a more participatory democracy. Their historical analysis shows that over the space of thirty years the opportunities to participate have indeed been widened and institutionalised. Since the nineteen eighties the market parties and non-governmental organisations have gradually gained more say in the decision-making process because they have started to carry the co-responsibility in a variety of interactive processes regarding the development and implementation of environmental policy.

Glasbergen & Goverde dealt with the ecological restructuring of democracy in Chapter 9. A reorientation on the relationship between the environment and politics in the nineteen nineties resulted in a debate on the 'Green Polder Model'. This was a new institution in environmental politics in which a variety of stakeholders became involved in the decision-making process regarding the quality of the physical environment. We are not only talking here about decision-making within the framework of environmental policy in the narrow sense, but also about decision-making on issues that can have radical consequences for the quality of the physical environment such as the infrastructure, agricultural development, urban renewal, etc.

The Green Polder Model encourages groups from a wide range of societal sectors, each with differing perspectives and interests, to develop a common policy practice for which they are all co-responsible. In the most far-reaching form of this model a certain authority would be designated to non-governmental parties to take up a part of the decision-making process on environmental issues in their own circle.

At the same time, the authors concerned here moderate the expectations for these new democratic structures. Leroy & Van Tatenhove state that new opportunities to participate have not for the time being led to a break-through in the existing power structures. Moreover, while the participation opportunities have been widened, it is still the same parties that make use of them. Glasbergen & Goverde argue that the Green Polder Model will only function properly if the balance of power between all parties is more or less equal and there is an awareness of their interdependency, not only now but in the long term as well.

12.5 The possibilities and limitations of the new method of governance

The new paradigm has become firmly established over the past few years. In fact, it has become the standard procedure in Dutch environmental policy. The main question dealt with in this section is what the new paradigm has produced. However, this is a very difficult question to answer. Firstly, the new paradigm has only been in use for a short time and the 'demonstrable' effects are difficult to track down. Secondly, it cannot be scientifically proven that the new paradigm works better than its predecessor did. After all, society is not a laboratory in which, under the same circumstances, two different experiments can be conducted and compared. In this sense the paradigm shift in environmental politics cannot be compared with Kuhn's paradigm theory. It is, however, quite possible to monitor the experiences and thus deduct the possibilities and limitations of the new method of governance. We shall base ourselves here on the chapters that dealt with production, consumption, the rural areas and the urban areas.

In Chapter 4 Vermeulen shows that with regard to businesses a very radical innovation has taken place in the methods of governance. Innovative forms of governance were gradually introduced, such as interactive management (including agreements entered into with branches of industry, environmental management) and self-regulation (including ISO standards). The production sector is in fact the perfect example of how the paradigm shift has had an effect in environmental policy practice. In this sector quite considerable reductions in emissions were achieved for several environmental themes such as waste, nuisance, acidification and soil pollution. The reduction in environmental impact caused by manufacturers must, according to Vermeulen, be attributed to forms of interactive governance. The collaboration between government, the knowledge institutes and a (limited) number of businesses was very important in this respect. Businesses benefit from this cooperation: they have an influence on the environmental goals, the time frames within which they must be achieved, and they get financial assistance for technological innovations. Industrial environmental policy occasionally goes hand in hand with an improvement in market position (national or international). However, organisations in civil society play only a modest role in the new governance method. This 'ecologisation of the economy' is apparently undertaken chiefly by a small leading group of – generally – internationally oriented companies. What today is referred to as green entrepreneurship is still a long way off for a large group of smaller businesses, argues Vermeulen. The most important challenge is apparently how to stimulate environmental ambitions (improving eco-efficiency by a factor of 4 or 10 in 25 years) and the new, associated procedures among a larger group of companies. Chapter 5 explains that this is all the more important given that governance through the demand side, the consumers, in order to reduce the environmental impact, will offer little solace for the time being. The consumer target group would seem to be unmanageable. Interactive governance is less suitable for this group: it is indefinable (the sum of a multitude of different individual interests) and badly organised.

There is also a paradigm shift in the way in which the environmental issues are tackled in rural areas. What was mainly a neo-corporativist method of governance has been replaced by cooperative governance strategies. In the neo-corporativist model, which for a long time had been dominant in policy on rural areas, there had been a staunch alliance between the government and the market, and civil society was sidetracked. In the new method of governance in rural areas the main aspect is the teamwork between the representatives of government, both national and local, businesses and non-governmental organisations in order to develop innovative solutions for the jointly perceived problems. Experience has been gained with these forms of governance in various projects. Driessen, Goverde & Van Tatenhove's contribution shows that positive results certainly can be achieved. It is possible to pursue a more 'customised' policy and all parties are becoming more aware of the problems involved and the new, particularly different options for tackling the problems. Generating support plays a major role in all projects and is indeed possible by way of interactive procedures. In terms of content, the new method of governance also implies that a predominant agricultural and economic perspective in rural area renewal has been replaced with a more integral approach in which issues of environmental protection, nature conservation and socio-economic development are seen as a whole. The authors also indicate that the expectations should not be exaggerated. Radical changes in rural areas cannot be achieved without radical reforms in agriculture. Interactive forms of governance are unable to make adequate contributions in this respect because developments in agriculture are steered mainly through national policies and policies of the European Union. The authors also reach the conclusion that new forms of governance can be more successful if more opportunities are offered for decentralisation and differentiation in the policy. This means a much weightier role for regions in shaping the policy.

Decentralisation is apparently a keyword in the new methods of governance in urban areas too (see Chapter 7). Here too do the innovations in governance methods emerge from the perceived limitations of hierarchical forms of governance. Meanwhile, policy on noise nuisance and soil cleaning have been decentralised from national government to the local and regional authorities. Simultaneously, the policy was also made more flexible. Instead of uniform environmental standards, different standards can be used. By doing this, specific circumstances can be taken more into account. Other aspects of renewal that have taken place in urban areas are similar to those that apply in rural areas: a more integral approach in which the environmental aspects are linked to aspects of spatial planning and economic development, plus more participation for citizens, businesses and non-governmental organisations in the decision-making process.

In summary, the total picture that has been created by the new paradigm is one of major changes in how specific environmental issues are being dealt with. Problems have been widened (by linking environmental issues to societal and economic issues), creativity in searching for solutions is stimulated, more parties are involved in the decision-making process, and there is apparently more support for the policy. Conversely, various actors also warn against the high expectations. The renewals in policy have not yet permeated sufficiently into all sectors of society and interactive governance is apparently hardly able to remedy major societal developments at national and international level. Van der Waals & Glasbergen quite rightly state in Chapter 7 that the importance of the changes in the organisation of policy with regard to how the environmental load will develop must to a certain extent be brought into perspective.

The new paradigm also involves several risks. Processes of interaction between the parties that participate in decision-making processes on environmental issues are sometimes less ideal than expected. Knowledge and information are used strategically in the negotiations, i.e. in a way which mainly benefits the party that owns specific sources of knowledge.

It would be naïve to assume that all available information in policy processes, in which major interests are at stake, can be exchanged openly as a matter of course. The power relations between the parties concerned can frustrate the learning processes.

Additionally, interactive processes in the initial stages also sometimes appear to lead to a widening of the problem definitions and options, but as soon as negotiations are started up, the perspective is radically narrowed down yet again. For instance, while 'main port development' aspects and the 'quality of the physical environment' were taken into broad consideration in a variety of interactive processes concerning the expansion plans of the Dutch national airport Schiphol, attention given by the parties concerned in the negotiations and in the political decision-making process were focused almost exclusively on the problem of noise nuisance. Also the interactive debates on the expansion of the international port of Rotterdam were originally broad in terms of scope, in which attention was given to a large number of aspects concerning the quality of the physical environment and socio-economic developments. Later, the negotiations were apparently limited to the construction of a large industrial estate and, to compensate, the construction of 750 ha of nature reserve and recreational land.

That interactive governance leads to innovative solutions, which also contribute towards a more sustainable society, has not been demonstrated convincingly to date. Satisfaction with regard to the process (support) often overshadows the quality of the solutions. In this respect we must immediately point out that this quality is also very difficult to determine. After all, as we set out above, it is impossible to apply two different forms of governance in two identical circumstances and then to compare the results (as is quite feasible in laboratory situations). Yet it is remarkable to see that the result in policy practice is mainly defended by pointing out the consensus, and to a lesser extent by pointing out the problems it has solved.

Another point we must make here concerns the participation of third parties in the process of developing policy and decision making. It is generally accepted that interactive governance, through a wider participation of third parties, leads to a more open policy process. And yet this is not always the case. Decision making in a process of interactive governance occurs in networks of public and private parties. These networks are composed on the basis of relationships of dependency between public and private actors. There is always a 'representation of interests' in these networks. Participation is neither matter of course nor a right, but is only feasible if all parties involved have reached agreement with regard to that participation. It is therefore possible for certain stakeholders to be excluded. In practice it appears that there are many degrees of interactive decision making. In some decision-making processes all sorts of interested parties (citizens, businesses, non-governmental organisations, administrators, experts, etc.) may participate, in other projects participation is reserved for a pre-selected group of interest groups. Sometimes participation only implies 'joining in the discussion', sometimes it results in 'co-production', i.e. the joint development of policy with shared responsibilities. For that matter, the participation of citizens and some (less professionally organised) non-governmental organisations is by no means free of all problems. Participation is a time consuming, often lengthy process, and the subject matter is for many people very complicated.

The last comment we must make here relates to the role played by the interest groups in interactive governance. For practical reasons the government prefers to consult with organisations that represent a large group of stakeholders. These organisations sometimes lack a democratic decision-making structure and culture. The result being that it is not always clear who or what it is they represent. For instance, who or what does an environmental organisation represent when it consults with government regarding the realisation of an important policy document on how the environmental problems are to be tackled? And are the members of that organisation also constantly informed of the standpoints taken and how the consultations conducted with the government are progressing? These are very important questions, particularly with a view to 'innovating democracy'.

12.6 Conclusions

In this book we have shown how the strategic views on the content of environmental policy and the procedures used have undergone a revolutionary change. We have called this change a 'paradigm shift'. According to the new paradigm, environmental policy can only be realised successfully if – in short – it is imbedded in a wider balancing up process in which social and economic interests are also taken into consideration, and if parties other than government parties are allowed to participate in the realisation and implementation of policy, such as businesses, non-governmental organisations and citizens. The changes we have pointed out have apparently also been observed – to a greater or lesser degree – in other countries. Jänicke & Weidner (1997, p. 311-312) for instance detected from a comparison of environmental policies of thirteen countries (spread over the entire world) "a remarkable increase in capacities for environmental policy and management in the last 25 years (...). With innovations such as new forms of participation, information rights, self-regulation, networking or cooperative planning, the ecological question has contributed to the general process of political modernisation. In more advanced countries, such as Denmark, Sweden, the Netherlands, Switzerland, Japan or Germany, it has also become a motor of economic modernisation and innovation". Lafferty & Meadowcroft (2001, p. 423-425) come to a similar conclusion for several European countries such as Norway, Sweden and the Netherlands. Nevertheless, they detect some differences in how the principle of sustainable development is implemented in the different countries.

The new paradigm has given significant stimulus in the Netherlands to the discussions on the greening of society. This not only concerns discussions on 'green production', 'green consumption' and the ecologically-sound structuring of rural and urban areas. Also greening democracy (i.e. the green polder model) is the subject of debate. The effects of these discussions and new procedures are perceptible. Firstly, the public debate on environmental issues has taken on a different character. Environmental issues are no longer seen as isolated but are now placed in a context of social and economic development, in a context of regional, national and international relationships, and in a context of taking action now with a view to the future. Secondly, the paradigm has led to a different division of responsibilities. Businesses and non-governmental organisations, in addition to national government, regional and local authorities, play a prominent role in the realisation and implementation of environmental policy. This can also be seen on different levels: regional, national and international. The new division of responsibilities leads to a more open decision-making process, a different sort of democracy, in which participation and cooperative decision making play a significant role.

The new paradigm has not, by the way, taken over the former one completely. The procedure that was a part of the former paradigm, i.e. a hierarchical governance, also still plays a role in environmental policy. Yet this procedure is highly modernised. Vermeulen gave several examples of this in Chapter 4: reinforcing and relieving the implementation system; the streamlining of procedures; new forms of flexible permits; making more use of general rules. In practice, this type of governance is used in a context of more interactive approaches. Legal rules on the one hand set out the limiting conditions within which these interactive approaches can be used. Additionally, they also play a part in the implementation of policy, for instance by creating guarantees for other interests or by binding the implementation of measures to certain rules.

We could state that the new paradigm fits in with the new societal relationships, i.e. network types of connections that have gradually arisen since the nineteen eighties. And yet the new paradigm is not without tension. Age-old certainties have elapsed, making way for uncertain relationships between the societal echelons. The government in particular is wrestling with its new role. On the one hand it will have to negotiate to an increasing extent with other parties on the policy to be pursued (one of many parties), on the other hand it has very special responsibilities (laid down in the Constitution) and must serve the general interest. In other words, it must simultaneously account for its responsibility within the policy networks within which it operates, and within the bodies of the representative democracy. This can give rise to tension in the pursuit of policy. After all, an agreement with the negotiating partners in a policy network is not automatically an agreement with parliament as well. It is particularly this field of tension that lies at the basis of the largest amount of resistance against the new paradigm. According to some people, the government looses its clout by interfering in all sorts of networks, and the primacy of politics is thus affected. Conversely, supporters of the new paradigm argue that the new procedure leads to faster and more effective implementation of policy and that democracy is thus enriched. It is the outcome of this debate that will determine the future of the new paradigm.

References

Glasbergen, P. (red.) (1994) *Milieubeleid, een beleidswetenschappelijke inleiding,* VUGA Uitgeverij, Den Haag, 1994, pp. 35-58.

Godfroij, A.J.A & N.J.M. Nelissen (red.) (1993) *Verschuivingen in de besturing van de samenleving,* Coutinho, Bussum.

Jänicke, M. & H. Weidner (eds.) (1997) *National Environmental Policies. A Comparitive Study of Capacity Building,* Springer Verlag, Berlin.

Kuhn, Th.S. (1962) *The Structure of Scientific Revolutions,* The University of Chicago Press, Chicago.

Lafferty, W.M. & J. Meadowcroft (eds) (2001) *Implementing Sustainable Development. Strategies and Initiatives in High Consumption Societies,* Oxford University Press, Oxford.

Leroy, P. (1994) 'De ontwikkeling van het milieubeleid en de milieubeleidstheorie', in: P. Glasbergen (red.) *Milieubeleid, een beleidswetenschappelijke inleiding,* VUGA Uitgeverij, Den Haag, 1994, pp. 35-58.

Snellen, I.Th.M. (1987) *Boeiend en geboeid; ambivalenties en ambities in de bestuurskunde,* Samsom Tjeenk Willink, Alphen aan den Rijn.

Teisman, G.R. (1992) *Complexe besluitvorming; een pluricentrisch perspektief op besluitvorming over ruimtelijke investeringen,* VUGA, Uitgeverij, Den Haag.

ENVIRONMENT & POLICY

1. Dutch Committee for Long-Term Environmental Policy: *The Environment: Towards a Sustainable Future.* 1994 ISBN 0-7923-2655-5; Pb 0-7923-2656-3
2. O. Kuik, P. Peters and N. Schrijver (eds.): *Joint Implementation to Curb Climate Change.* Legal and Economic Aspects. 1994 ISBN 0-7923-2825-6
3. C.J. Jepma (ed.): *The Feasibility of Joint Implementation.* 1995 ISBN 0-7923-3426-4
4. F.J. Dietz, H.R.J. Vollebergh and J.L. de Vries (eds.): *Environment, Incentives and the Common Market.* 1995 ISBN 0-7923-3602-X
5. J.F.Th. Schoute, P.A. Finke, F.R. Veeneklaas and H.P. Wolfert (eds.): *Scenario Studies for the Rural Environment.* 1995 ISBN 0-7923-3748-4
6. R.E. Munn, J.W.M. la Rivière and N. van Lookeren Campagne: *Policy Making in an Era of Global Environmental Change.* 1996 ISBN 0-7923-3872-3
7. F. Oosterhuis, F. Rubik and G. Scholl: *Product Policy in Europe: New Environmental Perspectives.* 1996 ISBN 0-7923-4078-7
8. J. Gupta: *The Climate Change Convention and Developing Countries: From Conflict to Consensus?* 1997 ISBN 0-7923-4577-0
9. M. Rolén, H. Sjöberg and U. Svedin (eds.): *International Governance on Environmental Issues.* 1997 ISBN 0-7923-4701-3
10. M.A. Ridley: *Lowering the Cost of Emission Reduction: Joint Implementation in the Framework Convention on Climate Change.* 1998 ISBN 0-7923-4914-8
11. G.J.I. Schrama (ed.): *Drinking Water Supply and Agricultural Pollution.* Preventive Action by the Water Supply Sector in the European Union and the United States. 1998 ISBN 0-7923-5104-5
12. P. Glasbergen: *Co-operative Environmental Governance: Public-Private Agreements as a Policy Strategy.* 1998 ISBN 0-7923-5148-7; Pb 0-7923-5149-5
13. P. Vellinga, F. Berkhout and J. Gupta (eds.): *Managing a Material World.* Perspectives in Industrial Ecology. 1998 ISBN 0-7923-5153-3; Pb 0-7923-5206-8
14. F.H.J.M. Coenen, D. Huitema and L.J. O'Toole, Jr. (eds.): *Participation and the Quality of Environmental Decision Making.* 1998 ISBN 0-7923-5264-5
15. D.M. Pugh and J.V. Tarazona (eds.): *Regulation for Chemical Safety in Europe: Analysis, Comment and Criticism.* 1998 ISBN 0-7923-5269-6
16. W. Østreng (ed.): *National Security and International Environmental Cooperation in the Arctic – the Case of the Northern Sea Route.* 1999 ISBN 0-7923-5528-8
17. S.V. Meijerink: *Conflict and Cooperation on the Scheldt River Basin.* A Case Study of Decision Making on International Scheldt Issues between 1967 and 1997. 1999 ISBN 0-7923-5650-0
18. M.A. Mohamed Salih: *Environmental Politics and Liberation in Contemporary Africa.* 1999 ISBN 0-7923-5650-0
19. C.J. Jepma and W. van der Gaast (eds.): *On the Compatibility of Flexible Instruments.* 1999 ISBN 0-7923-5728-0
20. M. Andersson: *Change and Continuity in Poland's Environmental Policy.* 1999 ISBN 0-7923-6051-6

ENVIRONMENT & POLICY

21. W. Kägi: *Economics of Climate Change: The Contribution of Forestry Projects.* 2000 ISBN 0-7923-6103-2
22. E. van der Voet, J.B. Guinée and H.A.U. de Haes (eds.): *Heavy Metals: A Problem Solved?* Methods and Models to Evaluate Policy Strategies for Heavy Metals. 2000 ISBN 0-7923-6192-X
23. G. Hønneland: *Coercive and Discursive Compliance Mechanisms in the Management of Natural Resourses.* A Case Study from the Barents Sea Fisheries. 2000 ISBN 0-7923-6243-8
24. J. van Tatenhove, B. Arts and P. Leroy (eds.): *Political Modernisation and the Environments.* The Renewal of Environmental Policy Arrangements. 2000 ISBN 0-7923-6312-4
25. G.K. Rosendal: *The Convention on Biological Diversity and Developing Countries.* 2000 ISBN 0-7923-6375-2
26. G.H. Vonkeman (ed.): *Sustainable Development of European Cities and Regions.* 2000 ISBN 0-7923-6423-6
27. J. Gupta and M. Grubb (eds.): *Climate Change and European Leadership.* A Sustainable Role for Europe? 2000 ISBN 0-7923-6466-X
28. D. Vidas (ed.): *Implementing the Environmental Protection Regime for the Antarctic.* 2000 ISBN 0-7923-6609-3; Pb 0-7923-6610-7
29. K. Eder and M. Kousis (eds.): *Environmental Politics in Southern Europe: Actors, Institutions and Discourses in a Europeanizing Society.* 2000 ISBN 0-7923-6753-7
30. R. Schwarze: *Law and Economics of International Climate Change Policy.* 2001 ISBN 0-7923-6800-2
31. M.J. Scoullos, G.H. Vonkeman, I. Thornton, and Z. Makuch: *Mercury - Cadmium-Lead: Handbook for Sustainable Heavy Metals Policy and Regulation.* 2001 ISBN 1-4020-0224-6
32. G. Sundqvist: *The Bedrock of Opinion.* Science, Technology and Society in the Siting of High-Level Nuclear Waste. 2002 ISBN 1-4020-0477-X
33. P.P.J. Driessen and P. Glasbergen (eds.): *Greening Society.* The Paradigm Shift in Dutch Environmental Politics. 2002 ISBN 1-4020-0652-7

KLUWER ACADEMIC PUBLISHERS – DORDRECHT / BOSTON / LONDON

MIX
Papier aus verantwortungsvollen Quellen
Paper from responsible sources
FSC® C105338

If you have any concerns about our products,
you can contact us on
ProductSafety@springernature.com

In case Publisher is established outside the EU,
the EU authorized representative is:
Springer Nature Customer Service Center GmbH
Europaplatz 3, 69115 Heidelberg, Germany

Printed by Libri Plureos GmbH
in Hamburg, Germany